教育部门职业教育与成人教育用书目录推荐教材
中等职业教育中餐烹饪专业课程改革新教材

西式厨房英语

吴海滨 主 编
黄 艳 孔 民 刘丽静 张 青 管 勇 副主编
李序锦 袁莉莉 李德成 郭文龙 参 编

北京师范大学出版社

图书在版编目（CIP）数据

西式厨房英语/吴海滨主编．—北京：北京师范大学出版社，
2018.8（2021.1重印）
中等职业教育旅游服务类专业课程改革新教材．烹饪
ISBN 978-7-303-24121-7

Ⅰ．①西… Ⅱ．①吴… Ⅲ．①西式菜肴－烹饪－英语－
中等专业学校－教材 Ⅳ．① TS972.118

中国版本图书馆 CIP 数据核字（2018）第 191503 号

营销中心电话 010-58802181 58805532
北师大出版社职业教育与教师教育分社网 http://zjfs.bnup.com
电子信箱 zhijiao@bnupg.com

出版发行：北京师范大学出版社 www.bnupg.com
　　　　　北京市西城区新街口外大街12-3号
　　　　　邮政编码：100088
印　　刷：天津市宝文印务有限公司
经　　销：全国新华书店
开　　本：787 mm×1092 mm 1/16
印　　张：9.5
字　　数：198 千字
版　　次：2018 年 8 月第 1 版
印　　次：2021 年 1 月第 2 次印刷
定　　价：33.50 元

策划编辑：易　新　　　　　责任编辑：赵媛媛
装帧设计：高　霞　　　　　美术编辑：高　霞
责任校对：韩兆涛　　　　　责任印制：陈　涛

版权所有 侵权必究

反盗版、侵权举报电话：010-58800697
北京读者服务部电话：010-58808104
外埠邮购电话：010-58808083
本书如有印装质量问题，请与印制管理部联系调换。
印制管理部电话：010-58808284

编委会

主　编： 吴海滨

副主编： 黄　艳　孔　民　刘丽静　张　青　管　勇

参　编： 李序锦　袁莉莉　李德成　郭文龙

近年来,我国职业教育事业快速发展。《国务院关于加快发展现代职业教育的决定》中强调,要坚持以服务发展为宗旨,以促进就业为导向,适应技术进步和生产方式变革以及社会公共服务的需要,加快现代职业教育体系建设,深化产教融合、校企合作,培养数以亿计的高素质劳动者和技术技能人才。职业教育发展和行业发展的新形势,为烹饪专业建设奠定了深厚的基础。"面向21世纪职业教育课程改革和教材建设规划"研究与开发项目,已将烹饪专业列为重点建设专业之一。

作为国家级重点职业学校,我校的烹饪专业一直是市级骨干专业,师生在市级、省级、国家级乃至国际比赛中屡获佳绩。在大力发展学校教育,积极探索教育改革模式的同时,我校还致力于加强国际教育交流与合作办学,把我校的优秀学生送出去,把国际烹饪专业的先进教育理念、技术人才请进来,培养了一大批懂专业、精技艺、会管理的优秀人才。在交流与合作的过程中,我们发现语言交流成为障碍。由于中西餐在原材料、计量单位、加工方法、烹饪理念等方面都有所不同,特别涉及一些专业术语时,学生很难准确表达。为此,我们编写了这本教材,以适应烹饪专业学生学习、对外交流和发展的需要。

本教材根据我校学生的专业特点和餐饮行业对从业人员的要求,在与烹饪专业教师以及校外行业专家共同调研和探讨后,由一线英语教师进行汇总并编写而成的。在此过程中,我们得到了青岛香格里拉大酒店、青岛威斯汀酒店、青岛酒店管理职业技术学院等企业、院校的大力协助。

本教材着眼于西式厨房英语,共分为五大模块,前四个模块涉及厨房概述、厨房用具、蔬菜水果、肉类原料,最后一个模块介绍了几道典型的西式早餐、主菜等。这五大模块内容,遵循了烹饪顺序,从用具到原料,到最后的完整食谱,由易到难,由浅入深,循序渐进,符合学生的认知规律,能够使学生更好地掌握学习内容。

本教材以生活化、情景化的方式引领学生走进烹饪英语的世界。教材以一位实习厨师的视角展开讲解,从工作面试开始到最后培训结束,共22天时间,每一天都有不同的内容要学习。这样的内容框架，能让即将进入实习岗位的学生有很强的代入感，也具有一定的趣味性。

  本教材知识性与情景化相结合，内容系统、层次清晰、实用性强，可作为与高等院校相关专业的衔接教材，也可作为各种烹饪专业英语培训班的教材，还非常适合酒店后厨从业人员、相关专业人员和英语爱好者自学使用。

  由于作者水平有限，书中难免存在疏漏之处，敬请批评指正。随着餐饮行业的发展，不断会有新的知识产生，我们会及时修订，合理增删，不断完善。

编 者

  下载资源：访问京师 E 课 http://zj.bnuic.com/mooc/ →注册用户→进入"京师职教"→在右上方下拉菜单中进入"我的工作台"→点击"融媒体课程"→点击"添加课程"→输入课程密钥"7fxunbxs"，即关联课程成功→进入"西式厨房英语"→点击"查看资源"，即可观看并下载相关资源。

目录 Contents

Chapter 1 Kitchen Overview — 1

- **Day 1** Get a New Job / 2
- **Day 2** Meet New Colleagues / 5
- **Day 3** New Working Place / 9
- **Day 4** Be Careful in Kitchen / 12

Chapter 2 Kitchen Tools — 15

- **Day 5** Knives / 16
  - Part 1 Chef's Knife / 16
  - Part 2 Paring Knife / 19
  - Part 3 Boning Knife / 22
  - Part 4 Palette Knife / 25
- **Day 6** Cookware / 28
  - Part 1 Frying Pan / 28
  - Part 2 Sauté Pan / 31
  - Part 3 Braising Pan / 34
  - Part 4 Stock Pot / 37
  - Part 5 Sauce Pan / 41
- **Day 7** Containers / 45
  - Part 1 Mixing Bowl / 45
  - Part 2 Sieve / 49
  - Part 3 Colander / 52

Day 8　Hand Tools / 55
　　Part 1　Whisk / 55
　　Part 2　Chopping Board / 58
　　Part 3　Grater / 61
Day 9　Deep Frier / 64

Chapter 3　Raw Materials—Vegetables and Fruits　　67

Day 10　Vegetables / 68
　　Part 1　Brassicas—Cabbage / 68
　　Part 2　Brassicas—Cauliflower / 70
　　Part 3　Stem and Shoots—Celery / 73
　　Part 4　Stem and Shoots—Asparagus / 75
Day 11　Vegetables / 77
　　Part 1　Tubers—Potato / 77
　　Part 2　Bulbs—Garlic / 79
　　Part 3　Bulbs—Onion / 81
　　Part 4　Roots—Carrot / 83
Day 12　Vegetables / 85
　　Part 1　Fruit Vegetables—Pumpkin / 85
　　Part 2　Fruit Vegetables—Tomato / 87
　　Part 3　Fruit Vegetables—Cucumber / 89
Day 13　Vegetables / 91
　　Part 1　Pods and Seeds—Peas / 91
　　Part 2　Leafy Vegetables—Lettuce / 93
Day 14　Vegetables / 95
　　Mushroom and Fungi—Mushroom / 95
Day 15　Fruits / 98

Chapter 4　Raw Materials—Meat　　　103

| Day 16 | Pan-fried Pork Chop / 104 |
| Day 17 | Pan-fried Breaded Chicken Cutlet / 107 |

Chapter 5　Kitchen Practice　　　111

Day 18	Bacon Omelette / 112
Day 19	Doughnuts / 116
Day 20	Shrimp Cocktail / 120
Day 21	Beef Vegetable Soup / 124
Day 22	Chocolate Ice Cream / 128

附录　厨房词汇及短语英汉对照表（按词性分类）/ 133

Chapter 1　Kitchen Overview

厨房概述

DAY 1
Get a New Job

▶▶ Today's Task

Hello! I'm Commis. Today I'll have an important job interview. Good luck to me!

Conversation

Commis: Good afternoon.

Chef: Good afternoon. Sit down, please. Do you have a resume?

Commis: Here you are.

Chef: What kind of job are you **applying for**?

Commis: I'm applying for a cook in western kitchen.

Chef: What kind of education have you had?

Commis: I **graduated from** Qingdao Cuisine Vocational School.

Chef: Do you have any **work experience**?

Commis: Yes, I once did a part-time job in a western kitchen.

Chef: OK. We'll contact you later.

Commis: Thank you for your consideration.

Chapter 1 Kitchen Overview

Notes: apply for   申请
graduate from   毕业
work experience   工作经历

▶▶ Words and Expressions

resume *n.* 个人简历	Resume

Name		Date of Birth	
Home Address		Contact Number	
...			

education   *n.*   教育

consideration   *n.*   考虑

a part-time job   一份兼职工作

contact   *v.*   联系

▶▶ Activity

Activity I   Finish the resume, then imitate an interview with your partner.

You study in Qingdao Cuisine Vocational School. Your major is western cooking. You are going to work in the hotel as a trainee. Now Shangri-la Hotel needs a commis in western kitchen. Please write an English resume.

Resume

Name		Date of Birth	
Home Address		Contact Number	
Email		Position	
Education			
Work Experience			

Activity Ⅱ Discussion.

Discuss with your group members and answer the following interview questions.

Why do you choose cuisine as your major?

What do you think of cooking?

What is your career goal?

What will you do to achieve your career goal?

厨房、餐饮和相关工作区域日常管理规定

1. 食物操作区和餐饮区内禁止吸烟。

2. 食物处理区不得饮食或嚼口香糖。

3. 不得吐痰、咬手指甲、抓或挖鼻孔。

4. 不得对着食品和饮料咳嗽、打喷嚏，在腹泻、呕吐或流感发烧时不可制作食品和饮料。

5. 不得接触或梳理头发。

6. 不得用手指或与食物接触的器具直接试味，只能使用匙羹试味，且试味的匙羹只能使用一次，如需再试，应换用另外一只。

7. 用餐、休息、如厕和培训时不得穿戴围裙，围裙需存放在指定地点。

8. 不得在制作食品和饮料时打电话。

DAY 2
Meet New Colleagues

▶▶ Today's Task

What an exciting day! I will meet lots of new colleagues. I'd better know their titles as soon as possible.

Commis: Good morning.

Chef: Good morning.

Commis: I'm new here. What do you do in the kitchen?

Chef: I'm the pastry chef.

Commis: What are you **in charge of**?

Chef: I am in charge of pastry kitchen.

Commis: Well, can you introduce other **titles** in this kitchen?

Chef: Sure. Mr Jones is our executive chef.

Commis: What is he in charge of?

Chef: He is in charge of the whole kitchen management. He often works in kitchen office.

Commis: I see. Any other titles?

Chef: Yes. When executive chef is off work, sous-chef will **take on** his **responsibility**. In the following, we have chef de partie, commis, trainee. According to different division of work, we have vegetable chef, pantry chef, butchery chef, broiler chef, pastry chef and so on.

Commis: And what next?

Chef: We have many cooks, such as sauce cook, fish cook, vegetable cook and so on.

Commis: I know. I am the commis. Haha. It happens to be the same as my name.

Chef: Exactly.

Notes: in charge of 负责，管理
title 头衔
take on 承担
responsibility 责任，职责

▶▶ Words and Expressions

executive chef 行政总厨

sous-chef *n.* 副厨师长

pastry chef 面点主厨

Activity

Activity Ⅰ Memorize different titles and translate them into Chinese.

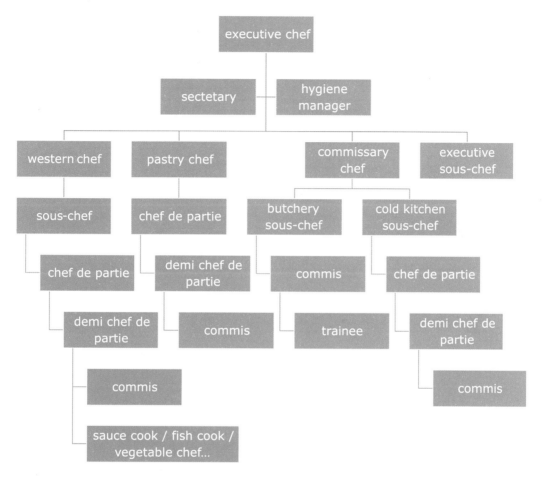

Activity Ⅱ Choose the suitable titles from the followings.

| butchery chef | commissary chef | western chef |
| executive chef | hygiene manager | pastry chef |

1. _____ is in charge of the whole kitchen.
2. _____ is in charge of making pastries.
3. _____ is in charge of making cold dishes.
4. _____ is in charge of cutting fish and meat.
5. _____ is in charge of hot western food.

Activity Ⅲ Please write down the names of the cook's uniform in English.

食品接触面的卫生及环境控制

1. 食品接触面的材质要求

与食品直接接触的表面应耐腐蚀、不生锈、光滑且易清洗，不宜采用木制、纤维或含有铁、铜等金属的材料。厨房内的工作台和案板（包括切肉、切菜），应尽量使用表面坚硬的材料（不锈钢或硬塑料材质），避免使用木制工作台和案板。

2. 食品接触面的维护和检查

与食品直接接触的设备和表面应制作精细，无粗糙焊缝、凹陷或破损，并始终保持完好，可以及时排水，无积存的食品残渣或积水。工作台和案板的表面应完整，无裂缝或明显缺陷，以避免裂缝中长时间积存食品残渣，发霉变质后成为污染源。

3. 食品接触面的清洗、消毒

食品接触面应定期进行清洗、消毒，大型设备每天使用后应进行彻底清洗、消毒，小型工具和台面应每次使用后进行清洗、消毒。

清洗、消毒时应彻底清除表面污垢，冲洗，并使用82℃以上热水浸泡或用含氯消毒剂喷洒消毒。不同厨房的工具应分开清洗、消毒。

Chapter 1　Kitchen Overview

DAY 3
New Working Place

▶▶ Today's Task

Today I should remember the kitchen plan as soon as possible.

Chef:　　Commis, this is the third day you work here. How is everything?

Commis: Great! I am very happy to **work with** you. My colleagues are all kind to me, but sometimes it takes me a long time to find them.

Chef:　　Don't worry. Here is our **kitchen plan**. Follow it and you'll get to know our working place soon.

Commis: Thank you.

Notes：work with　与……一起工作
　　　　　kitchen plan　厨房布局

Kitchen Plan

freezer	butchery section	fish section	cold kitchen	pastry section	beverage cooler		
kitchen store	chef's office					pick up area	restaurant
	scullery or wash-up section	vegetable preparation section		hot kitchen			

▶▶ Words and Expressions

cooler *n.* 冷饮，清凉饮料
         制冷装置，制冷机

scullery *n.* （洗碗盘、蔬菜等的）后厨房
            （炊具、食品等的）洗涤室

freezer *n.* 冰箱，冷藏库

beverage *n.* （汽水、茶酒等）饮料

Chapter 1　Kitchen Overview

▶▶ Activity

Activity　Do substitution work with your partner.
(vegetable chef / grill chef / fish chef / pastry chef / pantry chef / larder chef)
(vegetable preparation / hot kitchen / fish section / pastry / cold kitchen / butchery)

|  | Commis is talking about kitchen plan with Tom.
Example:
Commis: Where is <u>the executive chef</u>?
Tom:　　He is <u>in the chef's office</u>.
Commis: Where is <u>the grill chef</u>?
Tom:　　He is <u>in the hot kitchen</u>. |  |

餐饮业食品卫生安全之防虫防鼠

　　在厨房、加工车间和餐饮经营性建筑物内，禁止使用任何毒鼠药品或有毒诱饵，应使用机械式捕鼠器（鼠夹、捕鼠笼等）或物理性捕鼠器（胶性粘鼠板）。车间和库房进门两侧宜采用鼠夹，其余位置可每隔8~10米放置粘鼠板。厨房或建筑物的主要通道可放置鼠夹，建筑物外围放置捕鼠笼或捕鼠盒。

　　厨房内禁止将灭蝇灯直接悬挂于操作台面正上方。若车间内加工暴露的食品，则灭蝇灯应与食品间隔10米以上；若仅放置带包装的食品，则应间隔3米以上，以避免蚊蝇电击后飞溅误入食品中。

　　不论哪个季节，防虫防鼠设施都应保持24小时工作状态，切忌仅在生产时使用。

DAY 4

Be Careful in Kitchen

▶▶ Today's Task

Safety and hygiene in kitchen are very important. I should remember this.

Commis: This is chopping board.

Chef: Yes. We cut meat and other things on it with chef's knives.

Commis: Wow! So many kinds of knives. May I try it?

Chef: Sure. Make sure to use the knife **correctly** or it'll hurt your fingers.

Commis: All right. Here's a refrigerator.

Chef: Yes. We usually put food in it to keep it fresh. We also unfreeze meat in the refrigerator. Look! The door isn't **tightly** closed.

Commis: Let me close it, or food is easy to go bad.

Chef: Exactly! And it'll waste electricity. You must **pay attention to** it.

Commis: I will. What's this?

Chef: It's a deep frier. You should be careful. The oil temperature is too high. It can easily burn you.

Commis: It's really dangerous. Let's get away from it.

Chef: Don't worry. **As long as** you follow the instructions, it'll be safe.

Commis: I see. Shall we have a look at other equipment over there?

Chef: OK. This way, please.

Chapter 1   Kitchen Overview

Notes: correctly   正确地
tightly   紧紧地
pay attention to   关注，注意
as long as   只要

▶▶ Words and Expressions

chopping board 砧板	
chef's knives 主厨刀	
refrigerator *n.* 冷藏冰箱	
equipment *n.* 设备，设施	

▶▶ Activity

Activity Ⅰ   Answer the following questons.

1. What should a chef do when he uses knives?

2. If the door of the refrigerator wasn't tightly closed, what would happen?

3. How do you keep safe in the kitchen?

Activity II Translate the following words and expressions.

1. chef's knives _____
2. chopping board _____
3. refrigerator _____
4. deep frier _____
5. pay attention to _____
6. 正确用刀 _____
7. 把肉解冻 _____
8. 其他设施 _____
9. 关紧冰箱门 _____
10. 按指令去做 _____

Learning Tips

厨房区域个人卫生及个人财务管理规定

1. 制作食品时，所有员工不得戴假睫毛，化浓妆或喷香水。
2. 厨房员工不准佩戴手表、手镯、戒指等任何饰品。
3. 必须佩戴帽子，留有长发者，必须把头发扎在后面并覆盖。
4. 穿着整洁的工装、围裙，禁止在围裙上擦手。
5. 不得留长指甲，保持指甲卫生，不可涂抹指甲油（含无色）。
6. 笔和记事本不得放在口袋里，汤匙和小刀必须放在指定工作区域内。
7. 所有个人物品放在更衣室内，不得放在食品准备区内。

Chapter 2　Kitchen Tools

厨房用具

DAY 5
Knives

Part 1 Chef's Knife

▶▶ Today's Task

Today I'll show you common kitchen knives. They're very important for every cook.

Commis: Good morning, chef.

Chef: Good morning.

Commis: I want to chop some onions. What should I use?

Chef: Use your chef's knife. You can chop them, dice them or mince them.

Commis: All right. Can I use it to cut meat?

Chef: No, you can't. We use special knife for cutting meat.

Commis: I see. Thank you.

Chef: You're welcome.

Chapter 2　Kitchen Tools

▶▶ Words and Expressions

onion　*n.*　洋葱

chop　*v.* / dice　*v.* / mince　*v.*
切碎（粗碎）/ 切丁 / 切末（细碎）

use　*v.*　使用

▶▶ Activity

Activity Ⅰ　Answer the following questions according to the dialogue.

1. Is chef's knife for cutting vegetables?

2. What else can we do with a chef's knife?

Activity Ⅱ　Answer the followings with yes (Y) or no (N).

1. Can I use my chef's knife to cut vegetables?　　　　　　　　　　　(　　)
2. Can I use my chef's knife to cut meat without bones（骨头）?　　(　　)
3. Can I use my chef's knife to cut ribs（排骨）?　　　　　　　　　(　　)
4. Can I use my chef's knife to cut bread?　　　　　　　　　　　　　(　　)
5. Can I use my chef's knife to cut fish?　　　　　　　　　　　　　　(　　)

Activity Ⅲ　Do substitution work with your partner.

Commis:　I want to chop <u>onions</u> (potatoes / tomatoes / eggs / meat). What should I use?
Chef:　　Use your <u>chef's knife</u> (French knife).

Activity IV Pair work.

1. — Good morning, chef. I want to chop .
   What should I use?

   — Use your / Do it with your _____ .

2. — Good morning, chef. I want to dice .
   What should I use?

   — Use your / Do it with your _____ .

3. — Good morning, chef. I want to mince .
   What should I use?

   — Use your / Do it with your _____ .

4. — Good morning, chef. I want to chop .
   What should I use?

   — Use your / Do it with your _____ .

5. — Good morning, chef. I want to dice 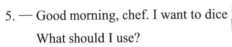 .
   What should I use?

   — Use your / Do it with your _____ .

Part 2　Paring Knife

▶▶ Today's Task

Today I'll show you common kitchen knives. They're very important for every cook.

Commis:　What's this?

Chef:　　It's a paring knife.

Commis:　What shall I do with it?

Chef:　　Use it to peel apples. You can also use a peeler.

▶▶ Words and Expressions

paring　*n.*　削皮，削下来的薄片	
paring knife　削皮刀，水果刀	
peel　*v.*　削，剥	

peeler *n.* 削皮器

▶▶ Activity

Activity Ⅰ Do substitution work with your partner.

Commis: What's this?

Chef: It's a paring knife (chef's knife / peeler).

Commis: What shall I do with it?

Chef: Use it to peel apples (chop onions / peel potatoes).

Activity Ⅱ Practice the dialogues with your partner.

1. — What's this?

    — It's a _____ .

    — What shall I do with it?

    — Use it to _____ .

2. — What's this?

    — It's a _____ .

    — What shall I do with it?

    — Use it to _____ .

3. — What's this?

    — It's a _____ .

    — What shall I do with it?

    — Use it to _____ .

如何正确握刀

正确握刀的手势是,将你的大拇指放在刀背和刀柄连接处的一边,用中指、无名指和小拇指自然地握住刀柄的另一边,食指则放在同侧刀背上、靠近刀柄的位置。用这种握刀方式切菜能让你的手腕在拥有对刀具最大控制力的同时承受最低限度的劳损。

切记不要把食指放在刀背顶上,这样会缩短刀的使用长度。尽管这看起来能带来一定的稳定性,但实际上会让刀具更易摇晃,而且会分散用刀的力量并降低用力的准确性。

Part 3　Boning Knife

▶▶ Today's Task

 Today I'll show you common kitchen knives. They're very important for every cook.

Commis: How sharp the knife is! Is it a special knife?

Chef:　　Yes.

Commis: What is it called in English?

Chef:　　It's a boning knife.

Commis: What is a boning knife for?

Chef:　　A boning knife is used to remove bones from meat.

Commis: Oh, I see.

▶▶ Words and Expressions

sharp　*a.*　锋利的，锐利的	
bone　*n.*　骨头	

Chapter 2　Kitchen Tools

boning knife　去骨刀

remove　*v.*　移动，取出

▶▶ Activity

Activity Ⅰ　Answer the following questions according to the dialogue.

1. Is boning knife a special knife?

2. What is a boning knife for?

Activity Ⅱ　Do substitution work with your partner.

Commis: What is it called in English?

Chef:　　It's a <u>boning knife</u> (chef's knife / paring knife).

Commis: What is a <u>boning knife</u> (chef's knife / paring knife) for?

Chef:　　A <u>boning knife</u> (chef's knife / paring knife) is used to <u>remove bones from meat</u> (chop vegetables / peel fruit).

Activity Ⅲ　Practice the dialogues with your partner.

1. — What is it called in English?

   — It's a 　　.

   — What is it for?

   — It is used to 　　.

2. — What is it called in English?

   — It's a 　　.

   — What is it for?

— It is used to .

3. — What is it called in English?

— It's a .

— What is it for?

— It is used to .

Learning Tips

常用物品消毒方法

刀具、案板：

使用200毫克/升的含氯消毒剂浸泡2分钟以上，并冲洗干净，待自然风干。

工作台面：

每日下班后，打开紫外线杀菌灯至少30分钟；或使用消毒剂对工作台面进行喷洒，待自然风干。

工作服：

使用82℃以上的热水浸泡至少5分钟，并清洗干净。

Part 4 Palette Knife

▶▶ Today's Task

 Today I'll show you common kitchen knives. They're very important for every cook.

Commis: Oh, that's a nice **birthday cake**. What shall we do next?
Chef: Let's put chocolate mousse on the cake.
Commis: With what?
Chef: With a palette knife.
Commis: OK, but I have never done this before.
Chef: It's easy.

Notes：birthday cake　生日蛋糕

▶▶ Words and Expressions

chocolate　*n.*　巧克力

mousse *n.* 慕斯	
palette *n.* 调色板	
palette knife 抹刀	

▶▶ Activity

Activity Ⅰ　**Answer the following questions according to the dialogue.**

1. What shall the commis and the chef do?

2. With what shall they put icing on the cake?

Activity Ⅱ　**Decide whether the followings are ture (T) or false (F).**

_____1. A boning knife is used to remove bones from meat.

_____2. A chef's knife can be used for many different jobs.

_____3. You can use a boning knife to cut big bones.

_____4. You can cut the bread with a palette knife.

_____5. A paring knife is used to peel fruits.

Activity Ⅲ　**Match the following knives with their English names.**

1. 　　　　　　　　　　A. chef's knife

2. 　　　　　　　　　　　　　　　　　　B. paring knife

3.  C. boning knife

4.  D. sharpening steel

5.  E. palette knife

Learning Tips

西餐餐具（一）

开胃食品餐叉：用于吃海鲜、开胃水果、龙虾，也可用于吃泡菜和橄榄。

沙拉餐叉：用于吃沙拉、鱼肉、馅饼、点心以及冷盘。

鱼餐叉：吃鱼时，鱼餐叉是用来替代正餐叉的。

正餐叉：用于除了鱼之外的所有主菜。

牛排刀：用于切熟肉，尤其是牛排。

DAY 6

Cookware

Part 1   Frying Pan

▶▶ Today's Task

Today, I will teach you the main usages of the common cookware.

Commis: Good morning, chef. Is this a frying pan?

Chef: Yes. But it is a special frying pan.

Commis: What is this frying pan for?

Chef: It is a small frying pan for crêpes.

Commis: What are crêpes?

Chef: They are French "baobing".

Commis: So a crêpe pan is a "baobing" pan.

Chef: Yes. **In Chinese** we call it a "baobing" pan.

Commis: I see.

Notes：in Chinese   用汉语，在汉语中

Chapter 2 Kitchen Tools

▶▶ **Words and Expressions**

fry *v.* 油煎	
pan *n.* 平底锅	
frying pan 带柄平底煎锅	
crêpe *n.* 小而极薄的烤饼（法）	
crêpe pan 用于烤薄饼的带柄平底锅	

▶▶ **Activity**

Activity Ⅰ   Answer the following questions according to the dialogue.

1. What are crêpes?

2. What is a crêpe pan for?

Activity Ⅱ   Practice the dialogues with your partner.

1. — What is it called in English?

   — We call it a  .

   — What is it for?

   — It's for   .

2. — What is it called in English?

 — We call it a   .

 — What is it for?

 — It's for   .

3. — What is it called in English?

 — We call it a   .

 — What is it for?

 — It's for   .

Activity Ⅲ Do substitution work with your partner.

Commis: Is this a <u>pan</u> (knife)?

Chef: Yes. But it is a special <u>pan</u> (knife).

Commis: What is it called in English?

Chef: We call it a <u>crêpe pan</u> (bread knife / palette knife).

Commis: What is the <u>crêpe pan</u> (bread knife / palette knife) for?

Chef: It's for <u>crêpes</u> (bread / cakes).

西餐餐具（二）

黄油刀：用于切小黄油块、软奶酪、酸辣酱以及开胃小菜。

正餐刀：用于除了鱼之外的所有主菜。

汤勺：用于甜点、麦片粥以及汤。

茶匙：用于咖啡、茶、水果以及某些甜点。

冰饮料勺：用于高脚杯中的饮料或甜点。

小咖啡勺：用于调味品、鱼子酱以及餐后咖啡。

Part 2　Sauté Pan

▶▶ Today's Task

Today, I will teach you the main usages of the common cookware.

Commis: Hello, chef. What do you call this pan?

Chef:　　It's a sauté pan.

Commis: What **are** you **going to do** with it?

Chef:　　I'll melt the butter.

Commis: And then?

Chef:　　I'll sauté potatoes in the pan.

Commis: Oh, I see.

Notes：be going to do sth. / will do sth.　打算做某事，将要做某事

▶▶ Words and Expressions

sauté　　*a.*　嫩煎的，（小火）慢炒的

西 式 厨 房 英 语

sauté pan   炒锅

melt   *v.*   熔化，溶解

butter   *n.*   黄油

▶▶ Activity

Activity Ⅰ   Answer the following questions according to the dialogue.

1. What is the chef going to do with the sauté pan?

2. What shall the chef sauté the potatoes with?

Activity Ⅱ   Practice the dialogues with your partner.

1. — What are you going to do?

   — I'll  with a   .

2. — What are you going to do?

   — I'll  with a   .

3. — What are you going to do?

   — I'll  with a   .

4. — What are you going to do?

   — I'll  with a   .

5. — What are you going to do?

   — I'll  with a _____ .

西餐知识

西餐厅一般比较宽敞，环境幽雅，就餐时便于交谈，因此，在公共关系宴请中，西餐是一种比较受欢迎又方便可取的招待形式。西餐源远流长，又十分注重礼仪，讲究规矩，所以，了解一些西餐方面的知识是十分必要的。一餐内容齐全的西餐主要包括：

第一，头盘，又称开胃菜；

第二，汤类（头菜）；

第三，蔬菜、冷菜或鱼（副菜）；

第四，主菜（肉食或面食）；

第五，餐后甜品；

第六，咖啡和茶。

Part 3 Braising Pan

▶▶ Today's Task

Today, I will teach you the main usages of the common cookware.

Commis: Good afternoon, chef. How do you braise beef?

Chef: First, we sear the meat in **hot fat**.

Commis: And then?

Chef: We cook it on the stove.

Commis: How?

Chef: In a braising pan. Watch. I will show you.

Notes: hot fat   热油

▶▶ Words and Expressions

braise   v.   （用文火）炖（肉等）

braising pan 炖锅，煨肉锅

sear *v.* 烙黄，灼色

stove *n.* 炉，灶

▶▶ Activity

Activity Ⅰ Answer the following questions according to the dialogue.

1. What cookware is mentioned in the dialogue?

2. How does the chef braise beef?

Activity Ⅱ Do substitution work with your partner.

Commis: Good afternoon, chef. How do you braise beef (deep-fry potatoes chips / sauté potatoes / peel an apple)?

Chef:　　In / With a braising pan (frying basket 油炸筐 / sauté pan / paring knife).

Activity Ⅲ Pracice the dialogue with your partner.

1. — How do you  ?

— In / With a  .

2. — How do you ?

— In / With a .

3. — How do you ?

— In / With a  .

4. — How do you ?

— With a .

Part 4　Stock Pot

▶▶ Today's Task

Today, I will teach you the main usages of the common cookware.

Commis: Excuse me, chef. Where shall I put the chicken?

Chef:　　Put all of the chicken in the stock pot.

Commis: What next?

Chef:　　Add enough water into the stock pot. Don't fill it **to the top**.

Commis: And then?

Chef:　　First, **bring the water to a boil**. Then lower the fire.

Commis: To simmer?

Chef:　　Right.

Notes：to the top　到顶部

　　　　bring the water to a boil　把水烧开

▶▶ Words and Expressions

stock　*n.*　高汤，原汤，汤料

西式厨房英语 ENGLISH FOR WESTERN COOKING

pot   *n.*   罐，壶

stock pot   （炖原汤用的）汤锅

add   *v.*   加入

fill   *v.*   装满，填满

boil   *n.*   沸点
      v.   煮沸

lower   *v.*   降低，减弱

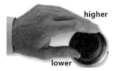

simmer   *v.*   煨炖，慢炖

▸▸ Activity

Activity Ⅰ Answer the following questions according to the dialogue.

1. Where shall the commis put the chicken?

2. How does the chef simmer the chicken?

Activity Ⅱ Do substitution work with your partner.

Commis: Excuse me, chef. Where shall I put the <u>chicken</u> (potatoes / eggs / beef)?
Chef: Put the <u>chicken</u> (potatoes / eggs / beef) in the <u>stock pot</u> (frying pan / sauté pan / braising pan).
Commis: OK.

Activity Ⅲ Practice the dialogues with your partner.

1. — Excuse me, chef. Where shall I put the ?

  — Put the in the .

  — OK.

2. — Excuse me, chef. Where shall I put the ?

  — Put the in the  .

  — OK.

3. — Excuse me, chef. Where shall I put the ?

  — Put the in the  .

  — OK.

4. — Excuse me, chef. Where shall I put the ?

— Put the  in the .

— OK.

西餐的分类

西欧式——又称"欧式",以英、法、德、意等国的菜品为代表,其特点是选料精纯、口味清淡,以款式多、制作精细而享有盛誉。

东欧式——又称"俄式",以俄罗斯为代表,其菜品特点是味道浓,油重,以咸、酸、甜、辣皆具而著称。

美式——以英国菜为基础的英国菜、法国菜、俄国菜、美国菜、意大利菜以及德国菜等。

Part 5　Sauce Pan

▶▶ Today's Task

Today, I will teach you the main usages of the common cookware.

Commis: Hi, chef. In what kind of pot can I cook this chicken?

Chef:　　Cook the chicken in a sauce pan.

Commis: What to do first?

Chef:　　Cook the chicken in butter.

Commis: Cook the chicken in butter in a sauce pan?

Chef:　　Right. Then we make chicken sauce.

Commis: In the same pan?

Chef:　　Right.

▶▶ Words and Expressions

sauce pan　（有盖、长柄的）深平底锅，炖锅，蒸煮锅　　

▶▶ Activity

Activity Ⅰ Answer the following questions according to the dialogue.

1. What tools should the commis use to cook the chicken?

2. What should the commis do first?

Activity Ⅱ Match the following pans or pots with their English names.

1. 　　　　A. sauté pan

2. 　　　　B. sauce pan

3. 　　　　C. stock pot

4. 　　　　D. braising pan

5. 　　　　E. frying pan

Activity Ⅲ Practice the dialogues with your partner.

1. — In what shall I cook the ?

   — Cook the in a .

   — What should I do first?

   — Cook the in .

Then fix the sauce to go with the .

2. — In what shall I cook the ?

— Cook the in a  .

— What should I do first?

— Cook the in .

Then fix the sauce to go with the .

3. — In what shall I cook the ?

— Cook the in a .

— What should I do first?

— Cook the in  .

Then fix the sauce to go with the .

法国饮食文化

　　饮食文化作为文化的一部分，往往也可以成为一面明镜。它与一个民族的历史传统有着千丝万缕、不可分割的联系。法国人一向以善于吃而闻名，法餐一直名列西餐之首，法国菜一向以精致、豪华的品位风靡于世。法餐与中餐相比，虽然在用料和口味变化上有所不及，但在餐具的考究、进餐环境的幽雅和隆重的餐桌仪式的细节等方面都非常出众。法餐严格选取各种新鲜原料，并不断创新烹饪方法。

Chapter 2　Kitchen Tools

DAY 7
Containers

Part 1　Mixing Bowl

▶▶ Today's Task

Today I'll show you some containers. They're not big, but they are very useful in the kitchen.

Conversation

Commis: Morning, chef. What are we going to cook today?

Chef:　　We'll make tuna salad.

Commis: OK. What kind of **container** should I use?

Chef:　　Use a mixing bowl.

Commis: Well, is this a mixing bowl?

Chef:　　Yes, it is.

Commis: Can I mix mayonnaise and tuna in this mixing bowl?

Chef:　　Sure. Go ahead.

Commis: Right. Thank you for your help.

Chef:　　You're welcome.

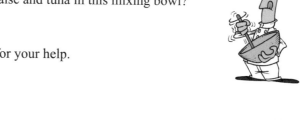

Notes：container　容器

西 式 厨 房 英 语 ENGLISH FOR WESTERN COOKING

▶▶ Words and Expressions

mix v. 混合在一起 n. 混合制品	
mayonnaise n. 蛋黄酱	
tuna n. 金枪鱼	
bowl n. 碗	
a mixing bowl 拌菜碗	

▶▶ Activity

Activity Ⅰ   Answer the following questions according to the dialogue.

1. What are the commis and the chef going to cook today?

2. What kind of container should they use?

Activity Ⅱ   Practice the dialogues with your partner.

1. — What's this?

  — It's a mixing bowl.

  — What can I do with the mixing bowl?

— Mix mayonnaise and in the mixing bowl.

2. — What's this?

   — It's a mixing bowl.

   — What can I do with the mixing bowl?

   — Mix mayonnaise and in the mixing bowl.

3. — What's this?

   — It's a mixing bowl.

   — What can I do with the mixing bowl?

   — Mix mayonnaise and in the mixing bowl.

4. — What's this?

   — It's a mixing bowl.

   — What can I do with the mixing bowl?

   — Mix mayonnaise and in the mixing bowl.

Activity Ⅲ Do substitution work with your partner.

(A)

Commis: What's this?

Chef: It's a mixing bowl.

Commis: What can I do with the mixing bowl?

Chef: Mix mayonnaise and <u>tuna</u> (potatoes / apples / tomatoes) in the mixing bowl.

(B)

Commis: What are we going to make today?

Chef: We'll make <u>tuna</u> (potato / fruit / vegetable) salad.

Commis: What kind of container should I use?

Chef: Use a mixing bowl.

西餐主菜服务标准程序

1. 检查客人点菜单，了解客人所点菜肴品名。
2. 检查并调整餐具，餐具应与所点菜肴相匹配。
3. 进厨房取菜：使用热菜盘；
   准备好调味汁；
   装托盘。
4. 为客人上菜：从客人右侧上菜；
   按女士、客人、主人的顺序上菜；
   向客人介绍菜名；
   上菜动作要轻。
5. 上调味汁：征求客人意见，介绍调味汁品种；
   从客人左侧上调味汁。
6. 祝客人用餐愉快，告退。随时了解客人的需求，按先女士，后男士，先客人，后主人的顺序提供服务。

Part 2　Sieve

▶▶ Today's Task

 Today I'll show you some containers. They're not big, but they are very useful in the kitchen.

Commis: Morning, chef. What is this called in English?

Chef:　　It is called a sieve.

Commis: Oh, what is a sieve for?

Chef:　　 It is for flour or breadcrumbs or icing sugar.

Commis: Shall I sift this flour with a sieve?

Chef:　　Sure. Go ahead.

Commis: Thank you.

▶▶ Words and Expressions

sieve　*a.*　筛，滤网
　　　　v.　筛，滤

sift　*v.*　筛，过滤，挑选

西 式 厨 房 英 语   ENGLISH FOR WESTERN COOKING

flour　*n.*　面粉，粉
　　　　v.　撒粉于……，把……做成粉

breadcrumbs　*n.*[*pl.*]　面包屑

sugar　*n.*　糖，白砂糖

icing sugar　糖粉

a flour sieve　面筛

▶▶ Activity

Activity Ⅰ　Answer the following questions according to the dialogue.
1. Can I sift flour with a sieve?

2. What is a sieve for?

Activity Ⅱ　Do substitution work with your partner.
Commis: What is this called in English?
Chef:　　It is called a <u>sieve</u> (mixing bowl / frying pan / paring knife).
Commis: What is <u>a sieve</u> (mixing bowl / frying pan / paring knife) for?
Chef:　　It is for sifting <u>flour</u> (making salad / frying food / peeling fruit).

Activity Ⅲ　Practice the dialogues with your parner.
1. — What is this called in English?

　　— It is called a  .

　　— What is it for?

— It is for  .

2. — What is this called in English?

— It is called a .

— What is it for?

— It is for .

3. — What is this called in English?

— It is called a .

— What is it for?

— It is for  .

4. — What is this called in English?

— It is called a .

— What is it for?

— It is for  .

Learning Tips

西餐汤的种类（Types of Western Soup）

西餐中的汤一般可以分为清汤和浓汤（茸汤）两大类。其中又有冷汤、热汤之分。清汤就是用牛肉、鸡肉、鱼及蔬菜等煮制的除去脂肪的汤，浓汤就是加入面粉、黄油、奶油等煮制的汤。

西餐汤还可以分为以下几种：清汤、奶油汤、蔬菜汤、浓（泥子）汤、冷汤、特制汤、地方性或传统性汤。

Part 3 Colander

▶▶ Today's Task

Today I'll show you some containers. They're not big, but they are very useful in the kitchen.

Commis: What shall I do with the carrots?

Chef: Are they **cooked**?

Commis: Yes.

Chef: Put them in a colander.

Commis: Why?

Chef: To drain them, of course.

Commis: And after I drain them?

Chef: Butter and salt the carrots.

Commis: **While** they are in the colander?

Chef: No. After you take them out.

Notes: cooked   煮熟的
         while   当……的时候

▶▶ Words and Expressions

carrot   *n.*   胡萝卜

Chapter 2　Kitchen Tools

colander　*n.*　（洗菜等用的）滤器

drain　*v.*　滴干，沥干

salt　*v.*　用盐腌，给……加盐
　　　n.　食盐

butter　*v.*　涂黄油

of course　当然

take...out　将……取出来

 Activity

Activity Ⅰ　Answer the following questions according to the dialogue.

1. Why does the commis put the carrots in a colander?

2. What shall the commis do with the carrots?

Activity Ⅱ　Do substitution work with your partner.

Commis: What shall I do with the carrots (flour / tuna), chef?
Chef:　　To drain (sieve / mix) them / it in / with a colander (flour sieve / mixing bowl).
Commis: OK.
Chef:　　Go ahead.

Activity Ⅲ　Practice the dialogues with your partner.

1. — What shall I do with the ?

— To  them with a .

— OK.

— Go ahead.

2. — What shall I do with the ?

— To them with a .

— OK.

— Go ahead.

3. — What shall I do with the ?

— To them with a .

— OK.

— Go ahead.

基础汤（Stock）

按照西餐特别是法国菜的烹饪习惯，要先做汤底，也称基础汤。基础汤是将含丰富蛋白质、胶质物的动物性原料放入锅中加水熬煮，使原料的营养成分溶于水中，成为营养丰富、味道鲜醇的汤汁。

西餐菜肴使用的基础汤一般有牛汤、鸡汤、鱼汤三大类。这三种汤汁分别是用牛肉和牛骨、老鸡和鸡架、鱼骨和鱼头做主料，并添加有香味的调味蔬菜（洋葱、胡萝卜、芹菜等）和香料（香叶、胡椒籽、百里香等）共同煮制而成的。煮汤的原料与水的比例应掌握好，水过多汤味不浓，水过少汤质易混。一般原料和水的比例是1∶3。比如，500克的牛肉和牛骨，放1500克左右的水。

制作时，采用弱火长时间煮沸，把原料的香味充分煮进汤汁中。煮制牛汤的肉毋须用高价肉，用牛颈、牛肩、牛胸、牛腿的肉就可以了；有时还可用其他边角料作为吊制汤的原料。如果吊制清汤，可将牛骨先放入烤炉内烤制一会儿，然后再放入清水中，这样味道更佳。

Chapter 2 Kitchen Tools

DAY 8
Hand Tools

Part 1 Whisk

▶▶ Today's Task

 Today I'll teach you how to use some hand tools. They are often used in the kitchen.

Conversation

Commis: Hello, chef. What is this machine?
Chef: It is a rotary whisk.
Commis: What shall I do with the whisk?
Chef: To beat eggs.
Commis: Shall I **beat** these eggs?
Chef: Yes, of course.
Commis: And then add oil?
Chef: First add vinegar, salt and lemon...and then add oil.
Commis: And beat the ingredients with a whisk?
Chef: Right. You will soon have mayonnaise.
Commis: That's great.

Notes: beat 打, 搅

Words and Expressions

whisk   *n.* （用金属丝盘绕而成的）打蛋器

oil   *n.* 油

vinegar   *n.* 醋

lemon   *n.* 柠檬，柠檬汁

ingredient   *n.* 烹调的原料，（混合物的）组成成分

rotary   *a.* 旋转的

rotary whisk   可转动的或手摇的打蛋器

and then   然后

Activity

Activity Ⅰ Answer the following questions according to the dialogue.

1. What shall the commis do with the whisk?

2. How does the commis make mayonnaise?

Activity II Do substitution work with your partner.

Commis: How shall I <u>beat the eggs</u> (drain the carrots / sift the flour / cut up the onions)?

Chef: You should use a <u>whisk</u> (colander / flour sieve / chef's knife).

Commis: Thank you.

Chef: You're welcome.

Activity III Pair work.

1. — Hello, chef. How shall I ?

   — You should use a .
   — Thank you.
   — You're welcome.

2. — Hello, chef. How shall I ?

   — You should use a .
   — Thank you.
   — You're welcome.

3. — Hello, chef. How shall I ?

   — You should use a .
   — Thank you.
   — You're welcome.

4. — Hello, chef. How shall I ?

   — You should use a .
   — Thank you.
   — You're welcome.

Part 2　Chopping Board

▶▶ Today's Task

Today I'll teach you how to use some hand tools. They are often used in the kitchen.

Chef:　　Commis, cut some onions and potatoes.

Commis: All right. But I need a chopping board to chop on.

Chef:　　Do you want a wooden chopping board or plastic one?

Commis: Which is better?

Chef:　　A plastic chopping board is better. Get it over there.

Commis: OK. How can I cut them, dices, julienne or batons?

Chef:　　Cut potatoes into batons. And cut onions into dices.

▶▶ Words and Expressions

board　*n.*　木板，板子	
julienne　*n.*　细丝 　　　　　*v.*　切成细丝	

Chapter 2　Kitchen Tools

plastic　*a.*　塑料的

dice　*n.*　丁
　　　v.　切成丁

baton　*n.*　条，粗条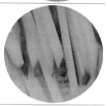

which　*pron.*　哪个，哪一个

better　*a.*　较好的，更好的（good 的比较级）

▶▶ Activity

Activity Ⅰ　Answer the following questions according to the dialogue.

1. What does the commis need to chop the onions on?

2. Which is better, the wooden chopping board or the plastic chopping board?

Activity Ⅱ　Match the following pictures with English.

1. 　　　　　A. cut...into batons

2. 　　　　　B. julienne...

3. 　　　　　C. slice...

4.    D. dice...

5.   E. shred...

西餐汤及配料（Western Soup and the Ingredients）

西餐汤风味别致，花色多样，世界各国都有其著名的有代表性的汤。例如，法国的洋葱汤、意大利的蔬菜汤、俄罗斯的罗宋汤、美国的奶油海鲜巧达汤、英国的牛茶配芝士条等。除了主料以外，我们还常常在汤面上放一些小料加以补充和装饰，如炸面包丁、蛋羹丁、菜丝、菜丁、奶酪、无味的饼干、荷兰芹、咸猪肉片（培根切片炒香）等。别看这些小配料，它们往往能起到画龙点睛的作用。

Part 3 Grater

▶▶ Today's Task

Today I'll teach you how to use some hand tools. They are often used in the kitchen.

Commis: Shall I grate the cheese?

Chef: Yes.

Commis: Where is the grater?

Chef: I'll find a grater for you…Here you are.

Commis: Thanks. I'll grate the cheese right now. We need some grated cheese.

Chef: Go right ahead.

▶▶ Words and Expressions

grater *n.* （将食物磨成碎块的）擦子，多功能刨刀（一种磨碎擦丝机械，可将干酪、蔬菜等磨成细丝或细末）	
grate *v.* 磨碎	

cheese *n.* 奶酪，干酪

grated cheese   磨碎了的奶酪

right now   立刻

go right ahead   去做吧（相当于 go ahead，但语气略强）

▶▶ Activity

Activity Ⅰ Answer the following questions according to the dialogue.

What should we use to grate the cheese?

Activity Ⅱ Do substitution work with your partner.

Commis: Where is the grater (mixing bowl / chef's knife / whisk)? I use it to grate the cheese (mix the vegetables / cut the onion / beat the eggs).

Chef: I'll find it for you. Here you are (Here's one).

Activity Ⅲ Practice the dialogues with your partner.

1. — Good afternoon, chef. Where is the ?

   I use it to .

   — I'll find it for you. Here you are.

2. — Good afternoon, chef. Where is the  ?

   I use it to .

— I'll find it for you. Here you are.

3. — Good afternoon, chef. Where is the ?

   I use it to .

   — I'll find it for you. Here you are.

4. — Good afternoon, chef. Where is the ?

   I use it to .

   — I'll find it for you. Here you are.

DAY 9

Deep-frier

▶▶ Today's Task

Today I'll show you a kind of equipment. It is used to deep-fry something.

Conversation

Commis: How shall I cook these potato chips?
Chef: Use the deep-frier.
Commis: What shall I do first?
Chef: Put the potato chips in a frying basket.
Commis: And then what?
Chef: Put the frying basket in the deep-frier.
Commis: Now?
Chef: Yes, now. We are making French fries.

▶▶ Words and Expressions

chip *n.* 条
      v. 把……切成条或片

deep-frier   深炸锅

fries   *n.*   油煎食品，油炸食品

▶▶ Activity

Activity Ⅰ   Do substitution work with your partner.

Commis: How shall I cook these potato chips (clean the oil / transfer the potato chips / drain the carrots / sift the icing sugar / cut the meat)?

Chef: Use the deep-frier (filter / spider / colander / sieve / chef's knife).

Activity Ⅱ   Translate the following sentences into English.

1. 把薯条放进油炸筐里。

   _____.

2. 把油炸筐放进深炸锅里。

   _____.

3. 我们正在做炸薯条。

   _____.

4. 桌子上有个油炸筐。

   _____.

5. 深炸锅要洗干净。

   _____.

Chapter 3　Raw Materials
—Vegetables and Fruits

食品原料之蔬果篇

DAY 10
Vegetables

Part 1 Brassicas—Cabbage

▶▶ Today's Task

Learn to prepare the cabbage.

Commis: How shall I cook the cabbage?
Chef: Cut each head of cabbage in half.
Commis: And then boil it?
Chef: That's right.
Commis: Shall I **add** salt **to** the water?
Chef: Yes. So the cabbage will keep green.
Commis: Are we going to stuff the cabbage?
Chef: Yes.
Commis: What dish do we make?
Chef: Stuffed cabbage with meat.
Commis: All right, chef.

Notes: add...to 加入……

Chapter 3　Raw Materials—Vegetables and Fruits

▶▶ Words and Expressions

cabbage　　*n.*　卷心菜

a head of　　一个，一颗……

▶▶ Activity

Activity Ⅰ　Do substitution work with your partner.

Commis: How shall I _____ (cook / boil / stuff) the cabbage?
Chef:　　Like this.

Activity Ⅱ　Make a conversation.

|  | Commis is talking about how to prepare the cabbage with Tom. |  |

卷心菜（Cabbage）

　　卷心菜又名洋包菜、包菜、圆白菜等，起源于地中海沿岸，16世纪传入中国，因其耐寒、抗病、适应性强、易储存、产量高等原因，在我国普遍种植，成为春、夏、秋季的主要蔬菜之一，是常用的烹饪食材。用英文表达"一颗卷心菜"为 a head of cabbage，"两颗卷心菜"为 two heads of cabbage。

Part 2 Brassicas—Cauliflower

▶▶ Today's Task

Learn to prepare the cauliflower.

Commis: Good morning, chef. What can I do for you?

Chef: Take a head of cauliflower.

Commis: OK. Shall I wash the cauliflower?

Chef: Yes, wash it well.

Commis: Right. Cauliflower is always dirty.

Chef: Soak the cauliflower in salt water.

Commis: For how many minutes?

Chef: For thirty minutes.

Commis: How shall I cook the cauliflower?

Chef: Boil it in water **with** lemon juice.

Commis: The lemon juice will keep it white.

Chef: That's right.

Notes: with 和……一起

Chapter 3 Raw Materials—Vegetables and Fruits

▶▶ Words and Expressions

cauliflower   *n.*   花菜，菜花

soak   *v.*   浸泡

dirty   *a.*   脏的

well   *ad.*   好，完全地，彻底地

keep   *v.*   保持，保存

▶▶ Activity

Activity Ⅰ   Do substitution work with your partner.

Commis: Shall I _____ (wash / soak / boil) the cauliflower?

Chef: Yes. / Sure. / Go right ahead. / Wait a minute. / Not right now.

Commis: _____ (For how long / How long / For how many minutes)?

Chef: For twenty seconds / For half an hour / For five seconds.

Activity Ⅱ   Complete the dialogue and role-play it.

Commis: Shall I _____ the cauliflower?

Chef: Yes, wash it _____.

Commis: Right. Cauliflower is _____ dirty.

Chef: _____ the cauliflower in salt water.

Commis: For how _____ minutes?

Chef: For thirty minutes.

Commis: How _____ I cook the cauliflower?

Chef: _____ it in water _____ lemon juice.

Commis: The lemon juice will _____ it white.

Chef: That's right.

花椰菜（Cauliflower）

花椰菜，又称花菜、菜花，是一种十字花科蔬菜，为甘蓝的变种。花椰菜的头部为白色花序，与西兰花的头部类似。花椰菜富含维生素B群、维生素C。这些成分是水溶性的，易受热溶出而流失，所以花椰菜不宜高温烹调，也不适合水煮。花椰菜原产地中海沿岸，其产品器官为洁白、短缩、肥嫩的花蕾、花枝、花轴等聚合而成的花球，是一种粗纤维含量少、品质鲜嫩、营养丰富、味道鲜美、人们喜食的蔬菜。

Part 3 Stem and Shoots—Celery

▶▶ Today's Task

Learn to prepare the celery.

Conversation

Commis: What do you call this in English?
Chef: It's celery.
Commis: What shall I do?
Chef: Wash it well.
Commis: And then?
Chef: Chop the celery.
Commis: For salad?
Chef: Yes. We will use the celery in the salad.
Commis: All right, chef.
Chef: Come on, boy.

▶▶ Words and Expressions

celery   *n.*   芹菜，西芹

Activity

Activity Ⅰ   Do substitution work with your partner.

Commis: What do you call this in English?

Chef: It's _____ (celery / cabbage / tomato).

Activity Ⅱ   Make a conversation.

 Commis is talking about how to prepare the celery with Tom.

西芹（Celery）

　　西芹原产于欧洲，茎段厚实，人们通常取其茎段食用。西芹可做生菜沙拉，也可加热烹调；因其爽脆的口感被大众所接受，近来也常被用于中式料理中。

Part 4 Stem and Shoots—Asparagus

▶▶ Today's Task

Learn to prepare the asparagus.

Commis: Nice day, isn't it?

Chef: Yes, boy.

Commis: What should we prepare today?

Chef: Asparagus.

Commis: Fresh or tinned?

Chef: Tinned.

Commis: All right.

Chef: Open this tin of asparagus.

Commis: Shall I reheat the asparagus?

Chef: Yes. Heat the asparagus.

Commis: In water?

Chef: Yes, in salt water.

ENGLISH FOR WESTERN COOKING

Words and Expressions

asparagus   *n.* 芦笋

reheat   *v.* 再加热

Activity

Activity Ⅰ   Do substitution work with your partner.

Commis: Open this tin of _____ (asparagus / tomatoes)?
Chef: Yes. Reheat them.

Activity Ⅱ   Make a conversation.

 | Commis is talking about how to prepare the asparagus with Tom. |

芦笋（Asparagus）

　　芦笋原产于欧洲，有白芦笋和绿芦笋两种。嫩茎未经日照即采收的为白芦笋（white asparagus），而经日照后变为绿色的为绿芦笋（green asparagus），市场上以绿芦笋较为常见，通常作为配菜或沙拉原料。白芦笋的经济价值非常高，常搭配高档食材烹调，味道鲜甜，如芦笋浓汤、煨白芦笋鱼子酱等。

DAY 11

Vegetables

Part 1 Tubers—Potato

▶▶ Today's Task

 Learn to prepare the potatoes.

Commis: Good afternoon, chef.

Chef: Good afternoon, boy.

Commis: What shall I do?

Chef: Get me a sack of potatoes.

Commis: Shall I wash them?

Chef: Yes. Clean the potatoes and peel them.

Commis: And after that?

Chef: Dice the potatoes.

Commis: Are we going to make potato soup?

Chef: Yes, that's right.

▶▶ Words and Expressions

potato   *n.*   土豆

▶▶ Activity

Activity Ⅰ   Do substitution work with your partner.

Commis: Shall I _____ (peel / chop) the potatoes?
Chef: Yes.

Activity Ⅱ   Make a conversation.

|  | Commis is talking about how to prepare the potatoes with Tom. |  |

> 马铃薯（Potato）
>
> 　　马铃薯又称洋芋，原产于南美洲。马铃薯是西式料理中主要淀粉质的来源之一，容易使人有饱足感。其烹调方法多样，如烤马铃薯、马铃薯饼、马铃薯泥等。

Chapter 3 Raw Materials—Vegetables and Fruits

Part 2 Bulbs—Garlic

▶▶ Today's Task

Learn to prepare the garlic.

Commis: Morning, chef. How are you today?

Chef: Fine, thanks. Let's prepare the garlic.

Commis: Shall I peel the garlic now?

Chef: Yes, please.

Commis: How shall I prepare the garlic?

Chef: Crush the garlic and then mince it.

Commis: And add the salt to the garlic?

Chef: That's right. We will mix the garlic with butter.

Commis: So we are making garlic butter today.

Chef: That's correct.

Words and Expressions

garlic   *n.*   蒜

crush   *v.*   压碎

Activity

Activity Ⅰ   **Do substitution work with your partner.**

Commis: Shall I _____ (peel / crush / grind / add) the garlic?
Chef: Yes. Go ahead.

Activity Ⅱ   **Make a conversation.**

	Commis is talking about how to prepare the garlic with Tom.	

大蒜（Garlic）

    大蒜又称蒜头，略带辛辣的口感；原产于中亚，为世界古老农作物之一。经过短时间日光曝晒后的大蒜，其风味更加醇厚，地中海菜系经常使用大蒜。

Part 3　Bulbs—Onion

▶▶ Today's Task

Learn to prepare onions.

Commis: What shall we prepare today?

Chef: 　　Onions.

Commis: Shall I peel the onions?

Chef: 　　Yes. With your cook's knife.

Commis: And then?

Chef: 　　Onions halved cut and thinly sliced.

Commis: Why?

Chef: 　　We need finely chopped onions for the soup.

Commis: What soup?

Chef: 　　French onion soup.

▶▶ Words and Expressions

onion   *n.*   洋葱

fine   *a.*   精致的，好的

done   *a.*   煮熟的

▶▶ Activity

Activity Ⅰ   **Do substitution work with your partner.**

Commis: Shall I _____ (peel / chop) the onions?

Chef: Yes.

Activity Ⅱ   **Make a conversation.**

	Commis is talking about how to prepare the onions with Tom.	

Part 4 Roots—Carrot

▶▶ Today's Task

Learn to prepare the carrots.

Commis: What wonderful carrots!

Chef: Yes, they are fresh.

Commis: How shall I cook them?

Chef: Wash the carrots carefully.

Commis: And peel them?

Chef: Yes. Cut the carrots into batons.

Commis: Shall I boil the carrots in salted water?

Chef: Yes. Remember to serve the carrots with butter.

Commis: Nice dish!

Chef: Good job, boy.

Commis: Thanks, chef.

▶▶ Words and Expressions

carrot   *n.*   胡萝卜

cut...into batons   把……切成条

▶▶ Activity

Activity Ⅰ   Do substitution work with your partner.

Commis: Cut the _____ (carrots / potatoes / asparagus) into batons?
Chef: Right.

Activity Ⅱ   Make a conversation.

|  | Commis is talking about how to prepare the carrots with Tom. |  |

Learning Tips

胡萝卜（Carrot）

　　胡萝卜，即一般所指的红萝卜，富含胡萝卜素及维生素A，在西餐料理中是不可或缺的主要蔬菜之一。胡萝卜可切割修饰出搭配主菜所需的形状作为配菜，或熬煮高汤、酱汁等。

Chapter 3 Raw Materials—Vegetables and Fruits

DAY 12

Vegetables

Part 1 Fruit Vegetables—Pumpkin

▶▶ Today's Task

Learn to prepare the pumpkins.

Conversation

Commis: Good afternooon, chef.

Chef: Good afternoon, boy.

Commis: What are we going to do?

Chef: We are going to make pumpkin cakes.

Commis: What shall we start with?

Chef: With the pumpkins.

Commis: Wash the pumpkins?

Chef: Yes.

Commis: What shall I do then?

Chef: Peel them and remove the seeds. Then dice the pumpkins.

Commis: OK.

Chef: Steam the pumpkins.

Commis: All right, chef.

西式厨房英语 ENGLISH FOR WESTERN COOKING

▶▶ Words and Expressions

pumpkin   *n.* 南瓜

steam   *v.* 蒸

▶▶ Activity

Activity Ⅰ   Do substitution work with your partner.

Commis: Steam _____ (potatoes / pumpkins)?
Chef: Right.

Activity Ⅱ   Make a conversation.

|  | Commis is talking about how to prepare the pumpkins with Tom. |  |

南瓜（Pumpkin）

　　南瓜适合生长于各种气候条件下，所以大多数国家都有南瓜的踪影。其原生品种来自于欧洲，西餐常用其煮汤、制作甜点，如圣诞节的代表甜点南瓜胡桃派。

Part 2 Fruit Vegetables—Tomato

▶▶ Today's Task

Learn to prepare the tomatoes.

Commis: What shall I do?

Chef: Wash the tomatoes.

Commis: Shall I cut out the centers?

Chef: Yes, please.

Commis: What next?

Chef: Put the tomatoes in boiling water.

Commis: For how long?

Chef: For fifteen seconds.

Commis: And then?

Chef: Put them immediately in ice-cold water.

Commis: Why?

Chef: To peel them easily.

Commis: I see. What shall we use the tomatoes for?

Chef: For tomato soup.

Words and Expressions

tomato *n.* 西红柿，番茄

center *n.* 蒂

Activity

Activity Ⅰ   Do substitution work with your partner.

Commis: Shall I _____ (cook / use / wash) the tomatoes?
Chef: Yes. Go ahead.

Activity Ⅱ   Make a conversation.

|  | Commis is talking about how to prepare the tomatoes with Tom. |  |

番茄（Tomato）

　　番茄品种很多，有本地生产的黑柿番茄、樱桃番茄、黄金小番茄、连株番茄；也有许多进口番茄，如牛番茄、白番茄、李子番茄等。番茄在西餐料理中有非常重要的地位，尤其是意大利料理。番茄的加工制品也很多，对西餐料理来说都非常重要，如番茄酱、番茄汁、番茄糊、浓缩番茄等。

Part 3 Fruit Vegetables—Cucumber

▶▶ Today's Task

Learn to prepare the cucumbers.

Commis: What shall I do now?
Chef: Wash the cucumbers.
Commis: And peel them?
Chef: That's right.
Commis: And then what to do?
Chef: Slice the cucumbers.
Commis: For salad?
Chef: Yes, for cucumber salad.
Commis: OK.

▶▶ Words and Expressions

cucumber   *n.*   黄瓜

▶▶ Activity

Activity Ⅰ Do substitution work with your partner.

Commis: Shall I _____ (cook / use / wash) the cucumbers?

Chef:　　Yes. Go ahead.

Activity Ⅱ Make a conversation.

|  | Commis is talking about how to prepare the cucumbers with Tom. |  |

配菜（Side Dishes）

　　配菜是西餐菜肴中不可缺少的组成部分。一般西餐做好后，还要在盘边或另一个盘内配上少量加工成熟的蔬菜或米饭、面食等，从而组成一道完整的菜肴。这种与主料相搭配的菜品叫配菜。

　　西红柿、西兰花、胡萝卜、紫甘蓝、圣女果、青辣椒、红辣椒、香菜以及黄瓜等都可以充当配菜。

DAY 13

Vegetables

Part 1 Pods and Seeds—Peas

▶▶ Today's Task

Learn to prepare the peas.

Commis: What shall we prepare today?
Chef: Peas.
Commis: Are the peas frozen or tinned?
Chef: Neither. We have fresh peas today.
Commis: Good. So we must shell them.
Chef: Correct. Shell them into a colander.
Commis: Which colander?
Chef: A colander with big holes.
Commis: Why?
Chef: The little peas will fall out of the bottom.
Commis: What are we going to do with the big peas?
Chef: Use them for stews.
Commis: And the little peas?
Chef: We will cook them. Then serve them with butter.
Commis: OK.

西式厨房英语 ENGLISH FOR WESTERN COOKING

▶▶ Words and Expressions

peas *n.* 豌豆	
stew *n.* 炖的汤，炖的菜	
frozen *a.* 冷冻的	

▶▶ Activity

Activity Ⅰ   Do substitution work with your partner.

Commis: Cut the _____ (beans / peas / leeks) into sections ?
Chef: Yes.

Activity Ⅱ   Make a conversation.

	Commis is talking about how to prepare the peas with Tom.	

豌豆（Peas）

　　豌豆最早生长于中东。市面上的豌豆有生鲜（从初夏到秋天都有）、冷冻、罐装和脱水等品种，可以做成泥糊，或者放在汤及煨菜中，也可以当成一般蔬菜烹调。入菜时，其外荚通常会被剥去，但也可以用来做豌豆汤。

Part 2　Leafy Vegetables—Lettuce

▶▶ Today's Task

Learn to prepare the lettuce.

Commis: Nice day. What vegetable should I prepare?

Chef:　　Lettuce.

Commis: How shall I wash the lettuce?

Chef:　　**Soak** the lettuce **in** water first.

Commis: For half an hour?

Chef:　　Yes.

Commis: Shall I cut the lettuce then?

Chef:　　No. We will use the whole leaves.

Commis: For salad?

Chef:　　Yes.

Commis: Oh, the salad is nice.

Notes：soak...in　浸泡

Words and Expressions

lettuce   *n.*   生菜，莴苣

whole   *n.*   整个，全部
          a.   完整的

Activity

Activity Ⅰ   **Do substitution work with your partner.**

Commis: Shall I cut the _____ (cabbage / potatoes / onions / carrots / tomatoes)?
Chef: Yes. Go ahead.

Activity Ⅱ   **Make a conversation.**

	Commis is talking about how to prepare the lettuce with Tom.	

结球莴苣（Iceberg Lettuce）

结球莴苣又称美生菜或西生菜，是最容易在市场上买到的生菜之一。莴苣品种很多，中西方料理皆有使用，西餐大多用于制作沙拉，中餐则用于热炒。

DAY 14

Vegetables

Mushroom and Fungi—Mushroom

▶▶ Today's Task

Learn to make mushroom stew.

Commis: Hello, chef. What are these called in English?

Chef: They are called mushrooms.

Commis: Oh! They are healthy food! What shall we do today?

Chef: We will make stew.

Commis: What do you want me to do?

Chef: Get me some mushrooms.

Commis: OK. And then?

Chef: Wash the sand on the surface of the mushrooms.

Commis: I see. What for?

Chef: The mushrooms are for the stew.

▶▶ Words and Expressions

mushroom   *n.*   蘑菇

brush   *n.*   刷子

healthy   *a.*   有益于健康的

▶▶ Activity

Activity Ⅰ   Do substitution work with your partner.

Commis: What are we going to do?

Chef: We'll make _____ (vegetable / ham and cheese / mushroom and cheese) stew.

Commis: What _____ (next / now / shall I do)?

Chef :   Mix the eggs with _____ (grated cheese / Swiss cheese / processed cheese).

Activity Ⅱ   Fill in the blanks and read the sentences aloud.

1. What are these called _____ English?

2. They are _____ food!

3. We will make _____ .

4. _____ do you want me to do?

5. _____ me some mushrooms.

6. _____ them _____ with a brush.

7. The mushrooms are _____ the stew.

Activity Ⅲ   Make a conversation.

	Commis is talking about how to make mushroom stew with Tom.	

蘑菇（Mushrooms）

在西餐中，蘑菇是经常用到的食材之一。蘑菇营养丰富，高蛋白、低脂肪，富含人体必需的氨基酸、矿物质、维生素和多糖等营养成分，味道鲜美。在烹饪中，常用的品种有牛肝菌（porcini mushroom）、蚝蘑（oyster mushroom）、食用小蘑菇（button mushroom）、羊肚菌（morel mushroom）、香菇（shiitake mushroom）、金针菇（enoki mushroom）、块菌（truffle）等。

DAY 15

Fruits

▶▶ Today's Task

Learn new words about fruits and know about how to make fruit salad.

Commis: Good morning, chef. What are we going to do today?

Chef: We'll make ice-cream salad.

Commis: What shall we start with?

Chef: Peel 2 bananas, 2 oranges, 1 apple and 2 kiwi.

Commis: And then?

Chef: Cut them into pieces.

Commis: What's the next?

Chef: Prepare 20 cherries and remove the seeds.

Commis: Yes, chef.

Chef: Put some ice-cream into a mixing bowl.

Commis: OK.

Chef: Add bananas, oranges, apples, kiwi and cherries into the mixing bowl.

Commis: And after that?

Chef: Mix them well and serve.

Commis: I see.

Chapter 3　Raw Materials—Vegetables and Fruits

▶▶ Words and Expressions

banana　*n.*　香蕉

orange　*n.*　橘子

apple　*n.*　苹果

kiwi　*n.*　猕猴桃

cherry　*n.*　樱桃

peach　*n.*　桃子

watermelon　*n.*　西瓜

greengage　*n.*　青梅子，青李子

plum   *n.*   李，梅

grape   *n.*   葡萄

pear   *n.*   梨

pineapple   *n.*   菠萝

blackberry   *n.*   黑莓

mango   *n.*   芒果

apricot   *n.*   杏

lychee   *n.*   荔枝

tangerine   *n.*   柑橘，红橘

melon *n.* 瓜，甜瓜

date *n.* 枣

lime *n.* 酸橙

fig *n.* 无花果

coconut *n.* 椰子

▶▶ Activity

Activity Ⅰ　Match the following pictures with the words.

kiwi　　　　mango　　　　orange　　　　cherry　　　　plum

grape　　　watermelon　　　lemon　　　lychee　　　pear

Activity II Tick off the odd word in each group.

1. peach   apple   greengage   pumpkin   lychee
2. orange   apricot   purple grape   blackberry   cabbage
3. pineapple pea   green bean   aubergine   string bean
4. celery   eggplant   tangerine   cauliflower asparagus
5. potato   coconut   watermelon   cherry   brapefruit
6. banana   mango   green grape   lemon   onion
7. melon   date   strawberry   pear   pepper
8. fig   lime   cantaloupe   persimmon   mushroom
9. salad   cucumber   tomato   lettuce   carrot

Activity III Translate the following salads into Chinese.

Caesar salad _____   ham salad _____

chicken salad _____   chicken-breast salad _____

egg salad _____   shredded chicken salad _____

fish salad _____   prawn salad _____

tomato salad _____   vegetable salad _____

crab salad _____   cucumber salad _____

冰激凌水果沙拉（Ice-cream Fruit Salad）

第一步：将香蕉、橘子、苹果、猕猴桃去籽，去皮，切块（大小随意），与红樱桃一起放入玻璃器皿中。

第二步：取冰激凌（多少随意，但多点好吃，最好能包住水果块），搅拌均匀即可食用。

Chapter 4 Raw Materials—Meat

食品原料之肉类

DAY 16
Pan-fried Pork Chop

▶▶ Today's Task

Learn to fry pork chop and describe the cooking procedures.

Conversation

Chef: **Cut** pig tenderloin **into** thick slices.
Commis: OK. What next? Add salt?
Chef: Yes, add salt, chicken powder and pepper.
Commis: And then?
Chef: **Blot** the pork dry, dredge in flour.
Commis: I've done it.
Chef: All right. Now **dip** in egg wash, and roll in breadcrumbs.
Commis: Shall I heat the pot now?
Chef: Yes. **Pour** the vegetable oil and **deep-fry** the pork chop.
Commis: How long should I deep-fry the pork chop?
Chef: Deep-fry it until golden brown and crisp.
Commis: What next? Can I serve it now?
Chef: No. You should serve with chips, pickles, lettuce, cooked peas and so on, and **sprinkle** with a little butter.

Notes: cut...into... 把……切成……
　　　　blot　吸干
　　　　dip　蘸
　　　　pour　倒
　　　　deep-fry　炸
　　　　sprinkle　淋上

▶▶ Words and Expressions

pork chop　猪排

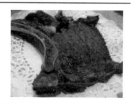

tenderloin　*n.*　腰部嫩肉，里脊肉

vegetable oil　植物油

▶▶ Activity

Activity Ⅰ　List materials used in fried pork chop.

1. ingredients: pig tenderloin _____ _____ _____ _____
2. condiments: _____ _____ _____
3. side dishes: _____ _____ _____ _____

Activity Ⅱ　Translation.

1. cut pig tenderloin into thick slices _____
2. add salt, chicken powder and pepper _____
3. dip with egg and breadcrumbs _____
4. heat the pot _____
5. deep-fry pork chop _____

6. 倒入花生油 _____

7. 把猪排炸至金黄 _____

8. 和小黄瓜一起上菜 _____

9. 淋上一点黄油 _____

10. 下一步做什么？_____

美国优质牛肉

美国特优级牛肉（USDA Prime）：这是牛肉评级中最高的等级，只有2%的牛肉能达到该等级。它纹理细腻，脂肪含量高（8%～11%）。

美国优级牛肉（USDA Choice）：这一等级的牛肉含有4%～7%的脂肪。在接受评级的牛肉中有超过一半可以被贴上该标签。

美国高级牛肉（USDA Select）：这一等级的牛肉含有3%～4%的脂肪。大约有三分之一的牛肉被评为该等级。这种牛肉比较瘦，因而在零售店比较畅销，但在饭店里通常属于等级最低的牛肉。

DAY 17

Pan-fried Breaded Chicken Cutlet

▶▶ Today's Task

Understand the recipe and practice making breaded chicken cutlet.

Commis: What are we going to do?

Chef: We'll make pan-fried breaded chicken cutlet.

Here is the recipe. Can you make this dish **according to** this recipe?

Commis: Well, it's really a big challenge. Let me **have a try**.

Chef: OK. Go ahead.

Pan-fried Breaded Chicken Cutlet (10 Servings)

Ingredients:

1.7 kg boneless chicken breast, cut into ten 170g portions;

1 g ground black pepper;

3 g salt;

3 large eggs, lightly beaten, or as needed;

340 g dried breadcrumbs;

142 g all-purpose flour, or as needed;

720 ml vegetable oil or clarified butter.

Procedures:

1. Blot the chicken breast dry, season with salt and pepper, dredge in flour, dip in egg liquid, and roll in breadcrumbs.

2. Heat about 3 minutes of fat about 177 ℃ in a large sauté pan over medium heat, working in batches. Add the breaded chicken to the hot oil and pan-fry on the presentation side for about 2 minutes, or until golden brown and crisp. Turn once and finish pan-frying on the second side until it reaches an internal tempetature 71 ℃, 1 or 2 minutes more.

3. Drain the chicken briefly on paper towels and serve immediately or hold hot for service.

Notes: according to 根据

have a try 尝试

▶▶ Words and Expressions

breaded chicken cutlet 面包糠炸鸡排	
breast *n.* 胸脯	
liquid *n.* 液体	
challenge *n.* 挑战	

▶▶ Activity

Activity Ⅰ   Please translate the recipe into Chinese.

1. Ingredients:

   _____

   _____

2. Cooking procedures:

   _____

   _____

   _____

   _____

Activity Ⅱ   Do substitution work and then read aloud.

Commis: What are we going to do?

Chef: We'll make <u>breaded chicken cutlet</u> (fried pork chop / fruit salad).

Commis: Can you make this dish according to <u>recipe</u> (procedures)?

Chef: Let me have a try.

畜肉类菜肴

　　畜肉类菜肴的原料取自牛、羊、猪等动物身上的各个部位，其中最受欢迎的、最有代表性的菜肴应数牛排。牛排类的主菜按肉的部位可分为许多品种，其中有西冷牛排（sirlion steak），T骨牛排（T-bone steak），肋眼牛排（ribeye steak）等，其他的猪肉菜肴、羊肉菜肴品种也很多。

　　肉菜常用的烹调方法有烤、煎、铁板、烩、焖等。

Chapter 5 Kitchen Practice

实训篇

DAY 18

Bacon Omelette

▶▶ Today's Task

Learn to make bacon omlette and describe the procedures.

美式早餐（American Breakfast）	
• Canned Juice or Fresh Juice	
peaches in Syrup	tomato juice
figs in Syrup	orange juice
pears in Syrup	apple juice
• Cereals（谷物类）	
corn flakes	rice krispies
musli	wheat
oatmeal	granola
• Eggs—Main Food	
fried eggs	boiled egg
poached eggs	scrambled eggs
omelet	
• Toast with Butter	
plain muffin	corn bread
croissant	pancakes
• Beverages	
white coffee	black coffee
black tea	green tea

Ingredients: clarified butter, cream, eggs.

Condiments: tomato, mushroom, onion, red bell pepper, green pepper, yellow pepper, Parmesan cheese powder, bacon, hash brown.

Procedures:

1. Shell 3 eggs in a mixing bowl.

2. Add cream.

3. Whisk eggs and season with salt and pepper.

4. Pour some clarified oil in a pan.

5. Add in onion, tomato, mushroom, red pepper, green pepper, yellow pepper and bacon.

6. Stir-fry them.

7. Pour in the eggs.

8. Stir and fold the eggs.

9. Place the omelette on the plate.

10. Garnish.

Commis: What are we going to do?

Chef: We'll make bacon omelette.

Commis: What shall we start with?

Chef: Shell 3 eggs in a mixing bowl.

Commis: And then?

Chef: Add cream seasoning, then mix them well.

Commis: That's right.

Chef: Pour some cooking oil in a pan.

Commis: I see.

Chef: Add in onion, tomato, mushroom, red pepper, green pepper, yellow pepper and bacon.

Commis: And after that?

Chef: Stir-fry them and pour in the eggs liquid.
Commis: And what next?
Chef: Stir and fold the eggs.
Commis: Shall I place the omelette on the plate?
Chef: Yes, and garnish the omelette with spring onions.

▶▶ Words and Expressions

omelette   *n.*   蛋卷

bacon   *n.*   培根

powder   *n.*   粉末

▶▶ Activity

Activity Ⅰ   Do substitution work with your partner.

Commis: What are we going to do?
Chef : We'll make _____ omelette (cheese / ham and cheese / mushroom and cheese).
Commis: What _____ (next / now / shall I do)?
Chef : Mix the eggs with _____ (grated cheese / Swiss cheese / processed cheese).

Activity Ⅱ   Complete the following sentences according to the dialogue and then read them aloud.

1. _____ 3 eggs in a mixing bowl.
2. _____ clear butter and whipping cream, then _____ them well.
3. _____ some cooking oil _____ a pan.

4. _____ them and _____ in the eggs.
5. _____ and fold the eggs.
6. _____ some salt and pepper powder on the omelette.
7. Shall I _____ the omelette on the plate ?
8. _____ the omelette with spring onions.

Activity Ⅲ   Make a conversation.

	Commis is talking about how to make omelette with Tom.	

与蛋相关的词汇

fried eggs   煎蛋

sunny side up   单面煎蛋

over easy   双面嫩煎

over hard   双面老煎

boiled egg   带壳水煮蛋

poached eggs   去壳水煮蛋

scrambled eggs   炒蛋

omelette   蛋卷

DAY 19
Doughnuts

▶▶ Today's Task

Learn to make doughnuts and describe the procedures.

欧陆式早餐（Continental Breakfast）	
• Fruit Juice	
orange juice	grapefruit juice
apple juice	
• Breads	
doughnuts	brioches（奶油蛋卷）
croissants（牛角面包）	
• Beverages	
tea	coffee
cocoa	
• Fruits	
bananas, apples, oranges and grapefruits	

Ingredients: baking powder, milk, all-purpose flour, sugar, salt, vanilla extract, egg, melted butter, vegetable oil.

Procedures:

1. Make dough with baking powder, milk, flour, sugar, salt, vanilla extract, eggs and melted butter.

2. Knead the dough smoothly and chill until cold, at least 1 hour or up to 3 hours.

3. Make doughnuts with a doughnut or cookie cutter.

4. Fry the doughnuts in the hot oil until golden, turning over once.

5. Drain on paper towels.

6. Dust with icing sugar while they are still warm.

Conversation

Commis: Shall we make doughnuts today?

Chef: Yes. It's classic food of continental breakfast.

Commis: What shall I do first?

Chef: Make dough with baking powder, milk, flour, sugar, salt, vanilla extract, eggs and melted butter.

Commis: And now?

Chef: Stir into all ingredients until well blended.

Commis: I've finished. And then?

Chef: Refrigerate it for an hour.

Commis: I see. (After an hour)

Chef: Now we'll make doughnuts with a doughnut or cookie cutter.

Commis: That's right. They look so lovely! What shall I do with the doughnut?

Chef: Fry the doughnut in the hot oil until golden, turning over once. Drain on paper towels.

Commis: That's OK.

Chef: Make the glaze with chocolate.

Commis: What shall I do then?

Chef: Dip the doughnuts into the glaze.

▶▶ Words and Expressions

doughnut   *n.*   甜甜圈

yeast   *n.*   酵母，酵母片

vanilla   *n.*   香子兰，香草

glaze   *n.*   釉，光滑面

knead   *v.*   揉合，揉捏

dip   *v.*   泡，浸，蘸

▶▶ Activity

Activity Ⅰ   Complete the following dialogues and then role-play them.

1. Commis: _____?
   Chef: Yes.
   Commis: _____?
   Chef: Make dough.
   Commis: _____?
   Chef: Knead the dough smoothly.
2. Commis: _____?
   Chef: Fry the doughnuts and cool them slightly.
   Commis: _____.
   Chef: Make the glaze with sugar.
   Commis: _____?
   Chef: Dunk the doughnuts into the glaze.

Chapter 5 Kitchen Practice

Activity Ⅱ Arrange the order of the statements according to the dialogue and then read them aloud.

1. _____ Fry the doughnuts and cool them.
2. _____ Make doughnuts with a doughnut or cookie cutter.
3. _____ Dunk the doughnuts into the glaze.
4. _____ Make dough with yeast, bread flour, sugar, salt, vanilla extract, eggs and melt butter.
5. _____ Make the glaze with sugar.
6. _____ Knead the dough smoothly and refrigerate it for at least 1 hour.

Activity Ⅲ Fill in the blanks with the words given below. Change the forms if necessary.

put cover break cut hold let cool constantly immediately

1. It's raining. Many people are _____ umbrellas (伞).
2. The box _____ when it fell.
3. Pour the soup into a heated tureen and serve _____.
4. Let the water _____ down slowly.
5. Stir the ingredients _____, please.
6. The chef _____ the meat into small pieces.
7. The chef said: "_____ the pan and cook for 2 hours."
8. You _____ too much salt in this food.
9. _____ me buy you a drink, please

Learning Tips

比萨饼（Pizza）

6 英寸比萨饼　small pizza

9 英寸比萨饼　regular pizza

12 英寸比萨饼　large pizza

铁盘比萨饼　pan pizza

手抛比萨饼　hand-tossed style pizza

DAY 20
Shrimp Cocktail

▶▶ Today's Task

Learn to make shrimp cocktail and describe the procedures.

Appetizer（开胃菜）	
shrimp cocktail	oyster cocktail
crab meat cocktail	chilled fruit cup
assorted relishes	russian black caviar
smoked salmon	smoked mackerel

Ingredients:

For the shrimp: seasoning, lemon, finely minced garlic, chili powder, salt, extra large tail-on raw shrimp.

For the cocktail sauce: tomato sauce, lemon juice, finely minced garlic, parsely.

Procedures:

1. Mix cocktail sauce with tomato sauce, lemon juice, finely minced garlic, parsley and chilli sauce.

2. Prepare a large bowl of ice water.

3. Add the seasoning, lemon, garlic, and salt to an 8-quart pot of water.

4. Boil them.

5. Add the shrimp to the pot and boil it to be bright pink.

6. Drain the shrimp and place them into the ice water for 2 minutes.

7. Peel the shrimp and serve with the cocktail sauce.

Chapter 5　Kitchen Practice

Commis: What are we going to do today?

Chef:　　Make shrimp cocktail.

Commis: What shall I do first?

Chef:　　Make cocktail sauce with tomato sauce, lemon juice, finely minced garlic, parsley and chilli sauce.

Commis: I finished. And then?

Chef:　　Prepare a large bowl of ice water.

Commis: That's right.

Chef:　　Add the seasoning, lemon, garlic, chili powder, and salt to a pot and boil them.

Commis: What shall I do with the shrimp?

Chef:　　Add the shrimp to the pot and boil it to be bright pink.

　　　　　(After a few minutes)

Commis: It's OK. And after that?

Chef:　　Drain the shrimp and place them into the ice water for 2 minutes.

Commis: Shall I peel the shrimp?

Chef:　　Yes. Peel the shrimp and serve them with the cocktail sauce.

Commis: That's wonderful!

Words and Expressions

shrimp cocktail 鸡尾冷虾

seasoning *n.* 调味品，佐料

shrimp *n.* 虾

minced *a.* 切碎的，切成末的

Activity

Activity Ⅰ Complete the following dialogue and then role-play it.

Commis: What are we going to do today?

Chef: _____.

Commis: What shall I do first?

Chef: _____.

Commis: I finished. And then?

Chef: _____.

Commis: That's right.

Chef: _____.

Commis: What shall I do with the shrimp?
Chef: _____.

Activity Ⅱ Fill in the blanks with the words given below, and then ask and answer the questions in pairs

| what | who | when | how long | how many | where | why |

1. _____ shall I do ?
2. _____ will make cocktail shrimp?
3. _____ cups of lemon juice do we need?
4. _____ shall I put the shrimp in the pot?
5. _____ shall I put the shrimp into the ice water?
6. _____ is the shrimp?
7. _____ don't you serve the shrimp with cocktail sauce?

DAY 21

Beef Vegetable Soup

▶▶ Today's Task

Learn to make beef vegetable soup and describe the procedures.

Soup（汤）	
beef vegetable soup	french onion soup
chicken mushroom soup	cream of tomato soup
Russian borsch	cream of mushroom soup
oxtail soup	

Ingredients: beef shank with bone, water, bay leaves, celery leaves, fresh parsley, salt, pepper powder, onion soup mix, stewed tomatoes, frozen vegetables, diced potatoes, sliced celery.

Procedures:

1. Remove any fat from the beef.

2. Simmer the beef shank in water with bay leaves, celery leaves, parsley, salt, pepper, and onion soup mix until the meat is tender.

3. Remove the bone, meat and bay leaves.

4. Cool broth and remove any fat.

5. Cut meat into bite size pieces and return to broth.

6. Add tomatoes, frozen vegetables, potatoes, and celery.

7. Cover and simmer for 25 minutes.

8. Serve the soup.

Conversation

Chef: The soup is very important in western cooking.

Commis: That' right.

Chef: Today we'll cook a typical soup "beef vegetable soup".

Commis: It's so exciting! What shall I do with the beef?

Chef: Remove the fat from the beef.

Commis: I have already done.

Chef: Simmer the beef bone with meat in water.

Commis: How about these ingredients?

Chef: Add the bay leaves, celery leaves, parsley, salt, and onion soup mix.

Commis: For how long?

Chef: Till the meat is tender. (After about 2 and a half hours)

Commis: What next?

Chef: Remove the bone, meat and bay leaves.

Commis: What shall I do with the broth?

Chef: Cool it and remove the fat.

Commis: And in the meantime?

Chef: Cut meat into bite size pieces and return to broth.

Commis: And what next?

Chef: Add tomatoes, frozen vegetables, potatoes, and celery.

Commis: OK.

Chef: Cover and simmer for 25 minutes.

Words and Expressions

beef *n.* 牛肉

broth *n.* 肉汤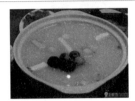

bay leaves 月桂叶，香叶

Activity

Activity Ⅰ Complete the following dialogues and then role-play them.

1. A: _____
   B: Remove the fat from the beef.
   A: I have already done.
   B: _____.
   A: _____?
   B: Add the bay leaves, celery leaves, parsley, salt, and onion soup mix.
   A: _____?
   B: Till the meat is tender.

2. A: _____?
   B: Remove the bone, meat and bay leaves.
   A: _____?
   B: Cool it and remove the fat.
   A: _____?
   B: Cut meat into bite size pieces and return to broth.

3. A: _____?
   B: Add tomatoes, frozen vegetables, potatoes, and celery.

A: OK.
B: _____.

Activity Ⅱ　Choose the odd word in each group.

1. (　) A. celery　　B. vegetable　C. lettuce　　D. cauliflower
2. (　) A. food　　　B. rice　　　　C. noodle　　D. bread
3. (　) A. broth　　 B. pork　　　 C. chicken　　D. beef
4. (　) A. bay　　　 B. thyme　　　C. spice　　　D. parsely
5. (　) A. halibut　 B. eel　　　　C. lobster　　D. seafood

Activity Ⅲ　Translate the following sentences into English.

1. ——杰克，早上好，我们一起去吃早饭好吗？
　——不，谢谢，我已经吃过了。

2. 那辆小汽车开进了宾馆的大门。

3. 离开家之前请你把门锁好。

4. 他拿着花，而她还在采更多的花。

5. 现在，在洋葱汤上撒些奶酪。

意大利面（Pasta）	
spaghetti　意大利实心面	penne　斜切管面
conghiglie　贝壳粉	fussili　螺旋粉
rigatoni　粗纹通心面	fettuccine　宽面
macaroni　通心粉	lasagne　千层面
linguini　黄油宽面	farfalle　蝴蝶粉
tortellini　意大利馄饨	ravioli　意大利饺子
cannelloni　意大利面卷	

DAY 22

Chocolate Ice Cream

▶▶ Today's Task

Learn to make chocolate ice cream and describe the procedures.

Typical Desserts	
chocolate ice cream	mango ice cream
chocolate sundae	custard pudding
vanilla cream cake	apple juice

Ingredients: cocoa powder, heavy cream, egg yolks, sugar, vanilla extract.

Procedures:

1. Heat the heavy cream in a medium saucepan.

2. Add the cocoa powder, stir constantly and simmer the mixture. Then remove it from the heat.

3. Whisk the egg yolks until they are bright in color.

4. Add sugar and whisk to combine.

5. Put the mixture to the saucepan and heat at over low heat.

6. Cook and stir frequently the mixture until it get thick.

7. Pour the mixture into a container and keep it at room temperature for 30 minutes.

8. Add in the vanilla extract.

9. Pour the mixture into an ice cream maker and refrigerate it for 4 to 8 hours.

Conversation

Chef: Do you like dessert?

Commis: Yes, I do.

Chef: We'll make chocolate ice cream today.

Commis: What shall we start with?

Chef: Heat the heavy cream in a medium saucepan.

Commis: And then?

Chef: Add the cocoa powder, stir constantly and simmer the mixture. And then remove it from the heat.

Commis: I see.

Chef: Whisk the egg yolks until they are bright in color.

Commis: And after that?

Chef: Add in sugar and whisk to combine.

Commis: And next?

Chef: Put the mixture to the saucepan and heat at over low heat.

Commis: What shall I do then?

Chef: Cook and stir it frequently until it gets thick.

Commis: That's OK now.

Chef: Pour the mixture into a container and keep it at room temperature for 30 minutes. (After 30 minutes)

Commis: Shall I add in the vanilla extract now?

Chef: Yes. Pour the mixture into an ice cream maker and refrigerate it for 4 to 8 hours.

Commis: OK, it must be wonderful!

Words and Expressions

whisk   *v.*   搅拌，搅动

gradually   *adv.*   逐步地，渐渐地

frequently   *adv.*   频繁地，经常地，时常，屡次

Activity

Activity Ⅰ   Complete the dialogue and role-play it.

Chef: _____?
Commis: Yes, I do.
Chef: _____.
Commis: _____?
Chef: Heat the heavy cream in a medium saucepan.
Commis: _____?
Chef: Add the cocoa powder and simmer the mixture.
Commis: _____.
Chef: Remove it from the heat.

Activity Ⅱ   Fill in the blanks with the proper forms of the verbs.

1. Now let's _____ (make) chocolate ice cream.
2. I _____ (make) a fruit salad now.
3. The commis _____ (refrigerate) the green beans an hour ago.
4. He _____ (mix) the ingredients and freeze them next.
5. _____ (place) the tuna in the center of the salad.
6. I've already _____ (chop up) the lettuce.

Activity Ⅲ   Fill in the blanks with the verbs according to the dialogue.

1. _____ the heavy cream ,add the cocoa powder and simmer the mixture.
2. _____ the egg yolks until they are bright in color.

3. _____ the mixture to the saucepan.

4. _____ in the vanilla extract.

5. _____ the mixture into the refrigerator to cool it for 4 to 8 hours.

Activity Ⅳ　Make a conversation.

 Commis is talking about how to make chocolate ice cream with Tom.

常与牛排搭配的酱汁

　　法国厨师总会在牛排旁加上许多配菜、酱汁，因为在他们看来，吃并不仅仅是为了吃（not eat for eat），而更多的是享受某种氛围。因此，菜的造型和摆放也成为了一种艺术，配菜、酱汁等每个细节都应力求达到完美。

　　不同的牛排要与不同的酱汁搭配。例如，西冷牛排（grilled beef sirloin）搭配的汁酱有：蘑菇汁（wild mushroom ragout）、青胡椒（creamy green pepper corn）、烧烤汁（classic BBQ）、鹅肝汁（foie gras butter）和法式伯那西酱汁（bearnaise）等。

附录 厨房词汇及短语英汉对照表（按词性分类）

名词 noun

厨房用具 utensils

boning knife	去骨刀
bone saw	切骨锯
bowl	碗
braising pan	炖锅，煨肉锅
carving knife	雕刻刀
chopping block	切肉墩
chopping board	切菜板
colander	滤器
conical strainer	圆锥形过滤器，过滤网，漏勺
chef's knife	厨刀
crêpe pan	用于烤薄饼的带柄平底锅
cutlet bat	拍肉板
deep frier	油炸锅
dish	碟
egg beater	打蛋器
electric bone saw	电动切骨锯
electric slicer	电动切片机
flour sieve	面筛
frying basket	油炸篮
frying pan	带柄平底煎锅
grater	擦子，礤床儿
grinder	磨碎机，绞肉机
kettle	壶，水壶
kitchen cutters	厨房刀具
ladle	长柄勺
liquidizer	榨汁器
mincer	切碎机，磨碎机，绞肉机
mixing bowl	拌菜碗
oven	烤箱，炉
oyster knife	开牡蛎刀

palette knife	铲刀
pallet knife	切蛋糕的刀
paring knife	水果刀
pizza cutter	比萨刀
plate	盘
poultry shears	家禽拔毛剪
pressure cooker	高压锅
roasting fork	烧烤叉
roasting pan	烧烤盘
roasting tray	烤盘
rolling pin	擀面杖
rotary whisk	可转动的或手摇的打蛋器
rubber spatula	橡胶刮铲
sauce pan	深平底锅，炖锅
sauté pan	煎锅
saw	锯
scissors	剪刀
serving spoon	公用匙
skewer	串肉针
skimmer	撇沫器
slotted spoon	漏眼匙
soup tureen	有盖汤碗
spider	漏勺
spoon	匙子，调羹
steak knife	顶端有锯齿的切牛排餐刀
steel	磨刀用的工具，钢钎
stew pan	（长柄）炖锅
stock pot	汤锅
tablespoon	汤匙，餐匙
teaspoon	茶匙
tray	托盘
whisk	打蛋器

食品原料 ingredient

蔬菜 vegetables

artichoke	洋蓟，朝鲜蓟
asparagus	芦笋
aubergine	茄子
bamboo shoot	竹笋

bean sprout	豆芽
beet	甜菜
bitter gourd	苦瓜
broccoli	西兰花
button mushroom	（食用）小蘑菇
cabbage	卷心菜，洋白菜
caraway	香菜籽
carrot	胡萝卜
cauliflower	菜花，花椰菜
celery	芹菜，西芹
coriander	香菜
corn	玉米
courgette	小胡瓜
cucumber	黄瓜
dried mushroom	干菇
eggplant	茄子
fungus	木耳
green bell pepper	青椒
green onion	青洋葱
green pea	青豆，青豌豆
leek	大葱
lettuce	莴苣，生菜
lotus root	藕
mushroom	蘑菇
enoki mushroom	金针菇
okra	秋葵
onion	洋葱
paprika	红柿椒
parsley	法国香菜，欧芹
pea	豌豆
potato	马铃薯，土豆
pumpkin	南瓜
spinach	菠菜
spring onion	春葱
string bean	四季豆
sweet potato	甘薯，甜薯
taro	芋头
tomato	番茄，西红柿
tomato sauce	西红柿汁（酱）
turnip	萝卜

white fungus	白木耳，银耳
white gourd	冬瓜
yam	山药
zucchini	西葫芦，荽瓜

水果 fruit

apple	苹果
apricot	杏
banana	香蕉
betelnut	槟榔
blackberry	黑莓
cantaloup	罗马甜瓜
cherry	樱桃
chestnut	栗子
coconut	椰子
durian	榴莲
fig	无花果
grape	葡萄
grapefruit	葡萄柚，西柚
greengage	青梅子，青李子
guava	番石榴
honeydew melon	密瓜，白兰瓜
juice peach	水蜜桃
kiwi	奇异果，弥猴桃
lime	酸橙
longan	龙眼
loquat	枇杷
lychee	荔枝
mandarin orange	橘
mango	芒果
melon	瓜，甜瓜
olive	橄榄
orange	橙，橘子
papaya	木瓜
peach	桃子
pear	梨子
persimmon	柿子
pineapple	凤梨，菠萝
plum	李子
rambutan	红毛丹

pomelo	柚子
star fruit	杨桃
strawberry	草莓
sugar-cane	甘蔗
tangerine	甘橘，红橘
water caltrop	菱角
water melon	西瓜
wax-apple	莲雾

肉禽海鲜类 poultry and seafood

abalone	鲍鱼
bacon	咸肉
beef	牛肉
chicken	鸡肉
cod	鳕鱼
crab	蟹
fried chicken	炸鸡
game	野味
goose	鹅
grouper	石斑鱼
grouse	松鸡
ham	火腿
hare	野兔
lobster	龙虾
mandarin	鳜鱼，桂鱼
meat	肉
medium	五分熟
mussel	贻贝，青口贝
mutton	羊肉
oyster	牡蛎
pheasant	山鸡
pork	猪肉
poultry	家禽
prawn	大虾，对虾
rare	三分熟
roast chicken	烤鸡
salmon	大马哈鱼
sardine	沙丁鱼
sausage	香肠
scallop	扇贝

shrimp	小虾
snail	蜗牛
steak	牛排
trout	鳟鱼
tuna	金枪鱼
turbot	比目鱼
turkey	火鸡
well done	全熟
white bait	银鱼

调味品 / 佐料 seasoning

anchovy	银鱼柳
anise	大茴香
apple butter	苹果酱
apricot jam	杏酱
barbeque sauce	烧烤酱
basil	罗勒
bay-leaf	香叶
black pepper	黑胡椒
blueberry jam	蓝莓酱
brown sugar	红塘
butter	黄油
butter fat	奶脂
caraway seed	小茴香
caviar	鱼子酱
cheese	芝士，奶酪
chili oil	辣椒油
Chinese pepper / flower pepper	花椒
cinnamon	肉桂，桂皮
clove	丁香
cream	奶油
cube sugar	方糖
cumin	孜然
curry	咖喱
curry powder	咖喱粉
dill	莳萝
fish sauce	鱼露
garlic	蒜
ginger	姜
gravy	肉汁

hoisin sauce	海鲜沙司，海鲜酱
honey	蜂蜜
plum sauce	梅子酱
maltose	麦芽糖
margarine	人造奶油
marjoram	牛膝草
marmalade	酸甜橙酱，马祖林
mayonnaise	蛋黄酱
milk	牛奶
mincemeat	水果馅
mint	薄荷
mustard	芥末
mustard oil	芥末油
nutmeg	肉豆蔻
olive oil	橄榄油
oregano	阿里根奴，牛至
oyster sauce	蚝油
palm oil	棕榈油
peanut oil	花生油
pepper	胡椒
rock sugar	冰糖
rosemary	迷迭香
saffron	藏红花
sage	鼠尾草
salad oil	色拉油
salt	盐
sesame oil	芝麻油，香油
soy sauce	酱油
soybean oil	豆油
sugar	糖
sunflower oil	葵花籽油
tarragon	他拉根香草
thyme	百里香
tomato ketchup / sauce	番茄沙司 / 酱
vanilla	香草精
vegetable oil	植物油
vinegar	醋
white pepper	白胡椒

动词 verb

add	加，增加
bake	烤
baste	在（烤肉、煎肉）上涂油，撒上（粉等）
beat	连续击打
blend	混合，把……混成一体
boil	达到沸点，煮沸，煮
braise	（用文火）炖（肉等）
braise	焖
bring	带来，拿来
butter	涂黄油
carve	切，切开，雕刻
check	检查，复查
chip	把……切成条或片
chop	砍，斩，剁劈
clean	把……弄干净
cook	烹调，煮，烧（食物）
cover	盖，覆盖
cover...with	用……盖上
crush	压碎，碾碎
cut	切，割
cut out	切掉，割掉
cut up	切开，割裂，弄伤，连根拔除，切碎
cut...into pieces	把……切成碎片
decorate	装饰
demonstrate	说明，演示
do with	处理
drain	滴干，沥干
fill	灌，倾注
filter	过滤
fix	放，固定
flatten	使平，变平，把……弄平
fold	折，折叠，合拢，包
fry	油炸，煎
garnish	装饰，加配菜于
go with	跟……相配
grate	磨碎，磨损
grill	烧，烤

grind	磨（碎），碾（碎）
gut	取出（肚肠）
heat	加热，加温
ice	使冰冷，加糖霜于（糕）上
insert	插入
keep	使得，使……保持
layer	层，分层
lift	举起，抬起
liquidize	使液化
lower	降低，使……低于，减弱
marinate	腌制
mash	捣碎，压碎
measure	量，测量，计量
melt	溶解，融解
mix	混合在一起，使混合，掺和
pare	削（果皮），刮，修
peel	剥皮，削皮
pick	捡，拿
poach	水煮
pour	倒，注，灌
press	压，榨取（汁等）
prick	刺（穿），扎（穿），刺小孔
pull out	拔出
reheat	再加热，对……重新加热
remove	移动，拿走，撤去，去掉，搬开
remove...from...	将……从……除去
reuse	再利用，重新使用
roast	烧烤，烤
roll	滚
salt	用盐腌，给……加盐
sauté	嫩炒
scramble	炒（蛋）
scrub	擦洗，擦净
sear	烙黄，烤焦
serve	上菜，开饭，伺候（顾客）
sharpen	磨快，磨尖，削尖
shell	去壳
shred	碎片，扯碎
sieve	筛，滤
sift	筛选，过滤

simmer	煨炖
skim	撇去（液体表面）的漂浮物
slice	切片，把……切成薄片，切下
soak	浸入，沁入
spill	洒，溢出
split	切开
sprinkle	撒落，撒上，撒
stack	堆，叠
sharp	磨，磨快
steam	蒸
stew	炖，煨，用文火煮
stick	插入，刺
stir	搅拌，搅动
stir-fry	煸炒
stuff	把……装满，填，塞
take out	拿出，取出
take...from...	将……从……拿出
take...out	将……取出
tear	撕，扯
tenderize	使软化
test	试验，试
toast	烤（面包）
touch	触摸
transfer	移动，移至
trim	整理，修剪
wash	洗
weigh	称重
wipe	擦，抹